Nitrogen-Containing Molecules: Natural and Synthetic Products Including Coordination Compounds

Nitrogen-Containing Molecules: Natural and Synthetic Products Including Coordination Compounds

Editors

Dimitris Matiadis
Eleftherios Halevas

MDPI • Basel • Beijing • Wuhan • Barcelona • Belgrade • Manchester • Tokyo • Cluj • Tianjin

Editors
Dimitris Matiadis
Institute of Biosciences and
Applications, National Centre
for Scientific Research
"Demokritos"
Greece

Eleftherios Halevas
Institute of Biosciences and
Applications, National Centre
for Scientific Research
"Demokritos"
Greece

Editorial Office
MDPI
St. Alban-Anlage 66
4052 Basel, Switzerland

This is a reprint of articles from the Special Issue published online in the open access journal *Molbank* (ISSN 1422-8599) (available at: http://www.mdpi.com).

For citation purposes, cite each article independently as indicated on the article page online and as indicated below:

LastName, A.A.; LastName, B.B.; LastName, C.C. Article Title. *Journal Name* **Year**, *Volume Number*, Page Range.

ISBN 978-3-0365-2257-9 (Hbk)
ISBN 978-3-0365-2258-6 (PDF)

Contents

About the Editors

Dimitris Matiadis received his PhD in organic chemistry at the National Technical University of Athens in 2013. Since 2018, he has worked in the National Centre for Scientific Research "Demokritos", in Athens, Greece as a postdoctoral researcher. He has participated in national research projects and received scholarships for postdoctoral research. In parallel, he teaches university undergraduate courses in the Department of Biomedical Sciences, School of Health and Care Sciences, at University of West Attica. He is a research grant evaluator for the Science Fund of the Republic of Serbia and reviewer for scientific journals of RSC, MDPI, Taylor & Francis and Wiley. The main focus of his research activities involves the following: synthesis of organic molecules and heterocycles, such as pyrazolines, tetramic acids and pyrazoles, characterization of organic and inorganic molecules by NMR and other spectroscopical methods and study, along with design of bioactive compounds as anticancer, antimicrobial or imaging agents.

Eleftherios Halevas received his PhD in bio-inorganic chemistry at the Department of Chemical Engineering, Aristotle University of Thessaloniki, in 2014. Since 2018, he has worked in the National Centre for Scientific Research "Demokritos", in Athens Greece, as a postdoctoral researcher. He has been actively involved in multidisciplinary national and international projects and has a long-standing experience and a strong publication record in the synthesis, crystallization, isolation and physico-chemical characterization of metal and lanthanide-complexes of flavonoids, curcuminoids and other natural and synthetic substrates. Furthermore, he has a strong scientific background in the field of bio-inorganic chemistry and nanocarrier synthesis. He is also experienced in single crystal X-Ray structure analysis. Additionally, he has a well-established role in the area of in vitro biological evaluation of various bioactive molecules and nanomaterials.

Editorial

Nitrogen-Containing Molecules: Natural and Synthetic Products including Coordination Compounds

Dimitris Matiadis * and Eleftherios Halevas *

Institute of Biosciences and Applications, National Centre for Scientific Research "Demokritos", 153 10 Athens, Greece
* Correspondence: matiadis@bio.demokritos.gr or dmatiadis@gmail.com (D.M.); lefterishalevas@gmail.com (E.H.); Tel.: +30-210-650-3558 (D.M. & E.H.)

Citation: Matiadis, D.; Halevas, E. Nitrogen-Containing Molecules: Natural and Synthetic Products including Coordination Compounds. *Molbank* **2021**, *2021*, M1282. https://doi.org/10.3390/M1282

Received: 7 September 2021
Accepted: 24 September 2021
Published: 27 September 2021

Nitrogen constitutes one of the most crucial elements in synthetic compounds, both in organic and in coordination chemistry. After all, urea is considered to be the first synthetic organic compound (it was first prepared in 1828 by Friedrich Wöhler from a mixture of the inorganic compounds silver cyanate ($AgOCN$) and ammonium chloride (NH_4Cl)).

From a medicinal chemistry viewpoint, nitrogen is a very common element in every major class of active pharmaceutical ingredients existing in heterocyclic and in acyclic molecules. Functional groups like amines, imines, nitriles, amides, and carbamates dominate the libraries of bioactive compounds, whereas the total number of unique drugs containing at least one nitrogen heterocycle among FDA-approved pharmaceuticals is 910 (84% among 1035 unique small-molecule drugs). The top five most frequent nitrogen heterocycles in this list are piperidine, pyridine, piperazine, β-lactam, and pyrrolidine [1,2].

Until today, the combined use of synthetic organic chemistry and coordination chemistry has generated a number of new and effective synthetic methods of high-yield product formation for important classes of coordination compounds of several nitrogen-containing ligands with numerous metal ions, giving rise to novel materials with exceptional physico-chemical, biological, medicinal, catalytic, and optical properties.

The aim of this Special Issue in the Molbank journal was, in principle, to collect original research articles related to one or more of the following: design and synthesis; structural characterization by means of NMR; X-ray crystallography; or other preliminary but significant results including, but not limited to, biological evaluation, theoretical calculations, molecular mechanics, and computational studies, material or physicochemical properties of nitrogen-containing molecules, and coordination compounds of nitrogen-containing molecules.

It was launched in August 2019 and, after two deadline extensions, was closed for new submissions in August 2021. Until this deadline, 16 exceptional contributions were made by authors from all over the world. Most of the contributions were made from authors based in Europe (12), with America (2) and Asia (2) following. Of note, is the broad geographic distribution of published articles, based upon the countries of the corresponding authors. Four of the published articles had authorship from Greece, two from Cyprus, two from the USA, and one from Russia, Japan, France, Saudi Arabia, Portugal, UK, and Spain. The majority (13) of the published content was in the field of synthetic chemistry, including synthetic methodologies and studies on chemical reactions and reactivity, while two papers dealt with coordination chemistry and one with structural elucidation. In particular, contributions related to organic synthesis may be divided in the subfields of aromatic reactivity (2), reactivity of heterocycles (2), synthesis of heterocycles (6), and detailed routine synthesis of very important compounds (3). Two of these papers included biological evaluation experiments.

The first two contributions were made from Kalogirou and Koutentis from Cyprus, in the field of aromatic reactivity. Kalogirou and Koutentis contributed with two communications on the reactivity of chlorinated pyrimidines. In their first paper, they focused on the displacement of carbon-2 or carbon-6 of polychlorinated pyrimidines using DABCO reagent (1,4-diazabicyclo[2.2.2]octane) with yields up to 52% [3]. The second publication demonstrates the synthesis of 2-cyanopyrimidines from the corresponding 2-thiomethyl pyrimidines starting from the 4,6-dichloro functionalized analogues. The synthetic route involves the "activation" of carbon-2 by displacing one or both chlorides with electron-donating methoxy groups [4].

Chronologically, the first article on reactivity of heterocycles was made by Fedorowicz and co-workers. The authors reported the synthesis of a bis-quinoline carboxylic acid derived from the corresponding isoxazolo-quinoline and dimethyl acetylenedicarboxylate (DMAD) under mild alkaline conditions. DFT calculations revealed that the product was more favorable than the double Michael addition isoxazolone product, which according to the proposed by the authors' mechanism, is formed as an intermediate [5].

In the second paper on this topic, the ring opening reaction of a bicyclic vinyl aziridine by an appropriate tetrazole-thiol was demonstrated by Fortunato et al. towards the synthesis of a novel tetrazole bearing cyclopentenol. The reaction occurred under mild conditions (37 °C, water) and according to authors' claims, in a regio- and stereospecific manner. The product can be readily further functionalized via its primary and secondary hydroxyl groups [6].

Arranged in order of publication date, the first contribution on the field of the synthesis of heterocyclic compounds or the cyclization towards to heterocyclic ring formation, was one of the two papers published by our group. In this study, a 4-ethoxycarbonylphenyl functionalized 2-pyrazoline was prepared, via the cyclization reaction between the corresponding bis(arylidene)acetone and ethyl hydrazinobenzoate in high yield and purity as evidenced by HPLC experiments. The compound was designed as an agent that would be active against resistant cancerous, since similar pyrazolines or pyrazole analogues have been reported in the literature or recent patents for this purpose [7,8]. However, though DNA thermal denaturation and DNA viscosity tests revealed that the compounds act as DNA intercalators, no activity against doxorubicin-resistant breast cancer or synergistic activity could be detected [9].

Konstantinova et al. reported the formation of a novel fused oxazinoxazine from ethyl 2-(hydroxyamino)propanoate and 2-aminophenol treated with S_2Cl_2. The structure of the product was also determined by X-ray crystallographic analysis. It is worth mentioning that this compound was obtained unexpectedly, instead of a dithiazole derivative and that following the same procedure using ethyl 2-oxopropanoate, the reaction did not occur [10].

Starting from β-ketoester derivative and 2-aminopyridine, Tenti and co-workers prepared a novel functionalized imidazo[1,2-*a*]pyridine-2(3*H*)-one in two steps. The first step involved the formation of a β-ketoamide, which was isolated and characterized. Next, its bromination in dichloromethane, following a cyclization by intramolecular S_N2 displacement of the intermediate, afforded the final nitrogen heterocycle. NMR experiments revealed that the enol isomer is predominant rather than the β-dicarbonyl analogue [11].

The synthesis of a spiro carbocyclic hydantoin with potential pharmacological interest was investigated by Pardali and co-workers in their contribution as a short note paper. The target molecule was obtained in three steps starting from 4-phenyl cyclohexanone. DFT calculations were performed in order to determine and study the most favorable structures [12].

The synthesis of a novel oxazolocoumarin incorporating an amine group was prepared by Vlachou et al. in three steps starting from 6-hydroxycoumarin. The first step involved the nitration of the starting material with cerium ammonium nitrate (CAN) resulting to the formation of three isolable compounds. The main product, 6-hydroxy-5,7-dinitrocoumarin, was cyclized with *p*-tolylmethanol via an Au/TiO$_2$ catalyzed reaction and subsequent reduction of the nitro group. The target compound was evaluated as possible antioxidant agent and inhibitor of soybean lipoxygenase with no or low activity [13].

The communication submitted by Ubeid et al. deals with the synthesis of a benzaldehyde compound derived from 2-mercaptobenzimidazole and its thiosemicarbazone analogue. The latter, which is a novel compound, was fully characterized and obtained in high purity and high yield (75%). However, the highlight of this study was the report of an improved method for the synthesis of the intermediate benzimidazole–benzaldehyde conjugate. According to the literature, this compound and similar compounds have been synthesized with the use of copper(I) iodide or irradiation and long reaction times (>12 h) [14,15], whereas the reported procedure involves the addition of K_2CO_3 for 1 h under reflux with a near to quantitative yield [16].

The following three papers involve the routine synthesis of one or more compounds of remarkable interest. All compounds were fully characterized and their synthetic protocols were described in detail. Asquith and Tizzard demonstrated the preparation and characterization of a 4-anilinoquinoline derivative, incorporating a difluoromethyl group, with potential biological properties. The target molecule was synthesized by routine synthesis from the corresponding 4-chloroquinoline and 3-(difluoromethyl)aniline. The crystallographic analysis of its hydrochloric acid is reported as well [17].

Husson and Guyard presented the synthesis of the novel 4′-(5-*N*-propylthiophen-2-yl)-2,2′:6′,2″-terpyridine, with potential application as a ligand for the preparation of metal complexes. The compound was obtained by the reaction between 2-acetylpyridine and 5-*N*-propylthiophene-2-carboxaldehyde, in high purity (>98%), as determined by quantitative NMR [18].

Banks et al. reported the synthesis of ten polar aromatic compounds bearing an alkynyl imine or amide. The products were obtained from inexpensive and commercially available starting materials in good yields under mild reaction conditions. The study is supplemented with a three-step synthetic protocol for the preparation of a catechol containing an alkynyl amine. This route proceeded via the protected catechol as acetonide [19].

Because there are no structurally characterized metal complexes of Schiff bases derived from *gamma*-amino acids, our research group was motivated to develop a novel Cu(II) complex based on a new Schiff base obtained by the condensation of *ortho*-vanillin with *gamma*-aminobutyric acid. The novel compounds were prepared in good yields and purity and were chemically, spectroscopically and structurally characterized through elemental analysis, HR-ESI-MS, FT-IR, UV-Vis, NMR, and single crystal X-ray diffraction. The crystal structure of the produced Cu(II) complex reflects an one-dimensional polymeric compound. The Cu(II) ion is bound to two singly deprotonated Schiff base bridging ligands forming a $Cu^{(II)}N_2O_4$ chelation environment, and a coordination sphere with a disordered octahedral geometry [20].

Takase et al. focused their interest on the development of transition metal complexes, particularly those in groups 7 and 8, bearing non-innocent azopyridyl ligands including 2,2′-azopyridine (apy) and 2-phenylazopyridine (pap), that find application as multifunctional materials with impressive magnetic properties. They reported a newly synthesized diradical neutral dicarbonylruthenium(II) complex, bearing two azopyridyl ligands coordinated as anion radicals, by introducing in its structure two CO molecules as ligand and reducing agents. Further magnetic studies on the complex revealed paramagnetic and antiferromagnetic interactions between the spins in each radical [21].

Aitken et al. reported the successful crystallization and determination of the X-ray structure of homopiperazine, a well-known heterocyclic compound with remarkable properties as a component of liquids for CO_2 capture and as a component of various organic and organic/inorganic supramolecular ionic salts and transition metal complexes. The X-ray studies revealed the pseudo-chair conformation of the molecule, with each NH acting both as a hydrogen-bonding donor and acceptor, leading to the formation of a complex network structure. Further NMR analysis facilitated the determination of the three one-bond CH coupling constants and the ^{15}N chemical shift [22].

This Special Issue on "Nitrogen-Containing Molecules: Natural and Synthetic Products Including Coordination Compounds" was launched two years ago with enthusiasm from two early-career researchers, based in the National Centre for Scientific Research "Demokritos", in Athens, Greece, with different backgrounds and expertise; organic and medicinal chemistry (Dr. Dimitris Matiadis) and inorganic and bioinorganic chemistry (Dr. Eleftherios Halevas). We are thrilled that the Special Issue has exceeded all our expectations with the submission and publication of many exceptional contributions from scientists on the highest level from all over the world. We would like to express our sincere gratitude to all authors who contributed and chose our Special Issue to publish their work. We also wish to thank all the reviewers for the evaluation of the submitted articles and the Editorial Office of Molbank, especially Mrs. Jade Lu, for the fast and professional handling of the manuscripts and for addressing any issue that arose in these two years. The success of this Special Issue shows the ongoing importance of this topic in chemistry and encourages us to consider editing other Special Issues in this area in the near future.

Author Contributions: Writing—original draft preparation, D.M. and E.H.; writing—review and editing, D.M. and E.H. All authors have read and agreed to the published version of the manuscript.

Funding: This research received no external funding.

Conflicts of Interest: The authors declare no conflict of interest.

References

1. Vitaku, E.; Smith, D.T.; Njardarson, J.T. Analysis of the Structural Diversity, Substitution Patterns, and Frequency of Nitrogen Heterocycles among U.S. FDA Approved Pharmaceuticals. *J. Med. Chem.* **2014**, *57*, 10257–10274. [CrossRef] [PubMed]
2. Heravi, M.M.; Zadsirjan, V. Prescribed drugs containing nitrogen heterocycles: An overview. *RSC Adv.* **2020**, *10*, 44247. [CrossRef]
3. Kalogirou, A.S.; Koutentis, P.A. Reactions of Polychlorinated Pyrimidines with DABCO. *Molbank* **2019**, *2019*, M1084. [CrossRef]
4. Kalogirou, A.S.; Koutentis, P.A. Synthesis of 2-Cyanopyrimidines. *Molbank* **2019**, *2019*, M1086. [CrossRef]
5. Fedorowicz, J.; Gzella, K.; Wiśniewska, P.; Sączewski, J. 2,2′-((1,4-Dimethoxy-1,4-dioxobutane-2,3-diylidene)bis(azanylylidene))bis(quinoline-3-carboxylic acid). *Molbank* **2019**, *2019*, M1093. [CrossRef]
6. Fortunato, M.A.G.; Siopa, F.; Afonso, C.A.M. (1R,4S,5S)-5-((3-Hydroxypropyl)amino)-4-((1-methyl-1H-tetrazol-5-yl)thio)cyclopent-2-en-1-ol. *Molbank* **2021**, *2021*, M1199. [CrossRef]
7. Manna, F.; Chimenti, F.; Fioravanti, R.; Bolasco, A.; Secci, D.; Chimenti, P.; Ferlini, C.; Scambia, G. Synthesis of some pyrazole derivatives and preliminary investigation of their affinity binding to P-glycoprotein. *Bioorg. Med. Chem. Lett.* **2005**, *15*, 4632–4635. [CrossRef]
8. Kolotova, E.S.; Shtil, A.A.; Novikov, F.N.; Chilov, G.G.; Stroganov, O.V.; Stroilov, V.S.; Zeifman, A.A.; Titov, I.Y.; Sagnou, M.; Alexiou, P. Novel Derivatives of 3,5-Divinyl-pyrazole for Medical Application. WO 2016/190770 A1, 1 December 2016.
9. Matiadis, D.; Mavroidi, B.; Panagiotopoulou, A.; Methenitis, C.; Pelecanou, M.; Sagnou, M. (E)-(1-(4-Ethoxycarbonylphenyl)-5-(3,4-dimethoxyphenyl)-3-(3,4-dimethoxystyryl)-2-pyrazoline: Synthesis, Characterization, DNA-Interaction, and Evaluation of Activity Against Drug-Resistant Cell Lines. *Molbank* **2020**, *2020*, M1114. [CrossRef]
10. Konstantinova, L.S.; Tolmachev, M.A.; Popov, V.V.; Rakitin, O.A. Ethyl 11a,12-Dihydrobenzo[b]benzo[5,6][1,4]oxazino[2,3-e][1,4]oxazine-5a(6H)-carboxylate. *Molbank* **2020**, *2020*, M1149. [CrossRef]
11. Tenti, G.; Cores, Á.; Ramos, M.T.; Menéndez, J.C. (E)-3-((2-Fluorophenyl)(hydroxy)methylene)imidazo[1,2-a]pyridin-2(3H)-one. *Molbank* **2021**, *2021*, M1212. [CrossRef]
12. Pardali, V.; Katsamakas, S.; Giannakopoulou, E.; Zoidis, G. 1-Methyl-8-phenyl-1,3-diazaspiro[4.5]decane-2,4-dione. *Molbank* **2021**, *2021*, M1228. [CrossRef]
13. Vlachou, E.-E.N.; Balalas, T.D.; Hadjipavlou-Litina, D.J.; Litinas, K.E. 4-Amino-2-(p-tolyl)-7H-chromeno[5,6-d]oxazol-7-one. *Molbank* **2021**, *2021*, M1237. [CrossRef]
14. Liu, B.; Lim, C.-H.; Miyake, G. Visible Light-Promoted C–S Cross-Coupling via Intermolecular Charge-Transfer. *J. Am. Chem. Soc.* **2017**, *139*, 13616–13619. [CrossRef]
15. Tan, B.Y.-H.; Teo, Y.-C. Ligand-free CuI-catalyzed chemoselective S-arylation of 2-mercaptobenzimidazole with aryl iodides. *Synlett* **2018**, *29*, 2056–2060. [CrossRef]
16. Ubeid, M.T.; Thabet, H.K.; Abu Shuheil, M.Y. Synthesis of 4-[(1H-Benzimidazol-2-yl)sulfanyl]benzaldehyde and 2-({4-[(1H-Benzimidazol-2-yl)sulfanyl]phenyl}methylidene)hydrazine-1-carbothioamide. *Molbank* **2021**, *2021*, M1273. [CrossRef]
17. Asquith, C.R.M.; Tizzard, G.J. 6-Bromo-N-(3-(difluoromethyl)phenyl)quinolin-4-amine. *Molbank* **2020**, *2020*, M1161. [CrossRef]
18. Husson, J.; Guyard, L. 4′-(5-N-Propylthiophen-2-yl)-2,2′:6′,2″-terpyridine. *Molbank* **2021**, *2021*, M1183. [CrossRef]
19. Banks, S.R.; Yoo, K.M.; Welker, M.E. Synthesis of Polar Aromatic Substituted Terminal Alkynes from Propargyl Amine. *Molbank* **2021**, *2021*, M1206. [CrossRef]
20. Halevas, E.; Hatzidimitriou, A.; Mavroidi, B.; Sagnou, M.; Pelecanou, M.; Matiadis, D. Synthesis and Structural Characterization of (E)-4-[(2-Hydroxy-3-methoxybenzylidene)amino]butanoic Acid and Its Novel Cu(II) Complex. *Molbank* **2021**, *2021*, M1179. [CrossRef]

21. Takase, T.; Kainuma, S.; Kanno, T.; Oyama, D. *Cis*-Bis(2,2′-Azopyridinido)dicarbonylruthenium(II). *Molbank* **2021**, *2021*, M1182. [CrossRef]
22. Aitken, R.A.; Sonecha, D.K.; Slawin, A.M.Z. Homopiperazine (Hexahydro-1,4-diazepine). *Molbank* **2021**, *2021*, M1200. [CrossRef]

Communication

Reactions of Polychlorinated Pyrimidines with DABCO

Andreas S. Kalogirou [1,*] and Panayiotis A. Koutentis [2]

[1]   Department of Life Sciences, School of Sciences, European University Cyprus, 6 Diogenis Str., Engomi, P. O. Box 22006, 1516 Nicosia, Cyprus
[2]   Department of Chemistry, University of Cyprus, P. O. Box 20537, 1678 Nicosia, Cyprus; koutenti@ucy.ac.cy
*   Correspondence: A.Kalogirou@euc.ac.cy; Tel.: +357-22892804

Received: 27 September 2019; Accepted: 12 October 2019; Published: 14 October 2019

Abstract: The reaction of 2,4,5,6-tetrachloropyrimidine (**4**) and 4,5,6-trichloropyrimidine-2-carbonitrile (**1**) with DABCO (1 equiv.), in MeCN, at ca. 20 °C gives 2,4,5-trichloro-6-[4-(2-chloroethyl)piperazin-1-yl]pyrimidine (**5**) and 4,5-dichloro-6-[4-(2-chloroethyl)piperazin-1-yl]pyrimidine-2-carbonitrile (**6**) in 42% and 52% yields, respectively. The new compounds were fully characterized.

Keywords: heterocycle; piperazine; pyrimidine; DABCO

---

1. Introduction

Piperazines and pyrimidines are useful nitrogen heterocycles owing to their use in pharmaceuticals. Among nitrogen heterocycles, these two rank as third and tenth in the most frequently used in U.S. FDA approved drugs [1]. Examples of piperazine-containing drugs include the antihypertensive prazosin and the antibiotic ciprofloxacin, while examples of pyrimidine drugs are fluorouracil (anticancer) and trimethoprim (antibacterial) (Figure 1).

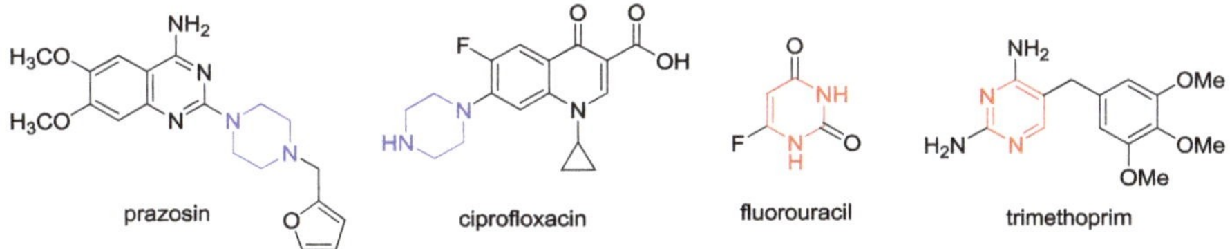

Figure 1. Piperazine- and pyrimidine-containing drugs.

Piperazines are often used as linkers in medicinal chemistry as well as to improve physicochemical properties of drug molecules such as water solubility and pharmacokinetic properties [2]. Unsymmetrical N-substituted piperazines and, in particular, those containing the *N*-ethylpiperazine moiety are useful pharmacophores but are often tricky to prepare [1–3]. One strategy to access these compounds is starting from the familiar tertiary amine 1,4-diazabicyclo[2.2.2]octane (DABCO). DABCO acts as a nucleophile in a variety of displacement reactions and often leads to the formation of quaternary ammonium salts that, in the presence of other nucleophiles, can ring open forming substituted *N*-ethylpiperazines [2–5].

Of particular interest are *N*-(2-chloroethyl)piperazines as these can be further functionalized via the 2-chloroethyl group. Surprisingly few reports of such compounds are found in the literature [6–11], and often the chloroethyl moiety was not isolated but converted in situ to other derivatives by nucleophilic displacement of the chloride [2–5].

As part of our ongoing work in the chemistry of 1,2,6-thiadiazines [12,13], we identified 4,5,6-trichloropyrimidine-2-carbonitrile (**1**) as a product of the chloride-induced thermal degradation of 3,4,4,5-tetrachloro-4*H*-1,2,6-thiadiazine (**2**) (Scheme 1) [12], while the same product has reappeared in previous work with 1,2,6-thiadiazines [13–15].

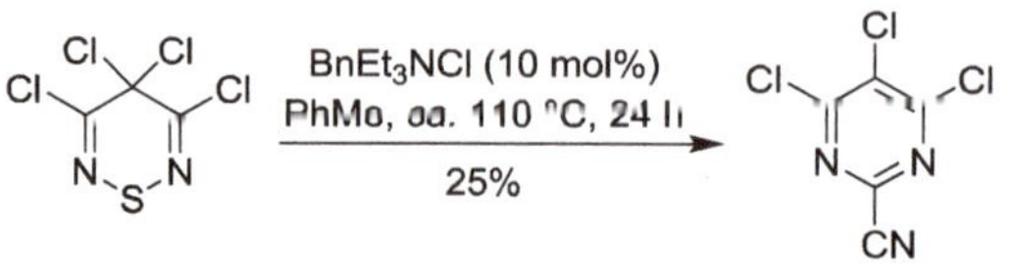

Scheme 1. Isolation of trichloropyrimidine **1** from 3,4,4,5-tetrachloro-4*H*-1,2,6-thiadiazine (**2**).

We are interested in studying the use of trichloropyrimidine **1** as a synthetic scaffold as it offers multiple sites of reactivity towards heteroatom nucleophiles or organometallic reagents. Previous efforts to access pyrimidine **1** involve the use of the starting material 4,6-dichloro-2-(methylthio)-pyrimidine (**3**) [16,17]. Another potentially useful scaffold for accessing pyrimidine **1** is the readily available 2,4,5,6-tetrachloropyrimidine (**4**), prepared by the treatment of barbituric acid with a refluxing mixture of PCl_5 and $POCl_3$ in 67% yield [18]. Retrosynthetically, the C2 cyano group of pyrimidine **1** could be introduced via a nucleophilic displacement of the C2 chloride of tetrachloropyrimidine **4** (Scheme 2).

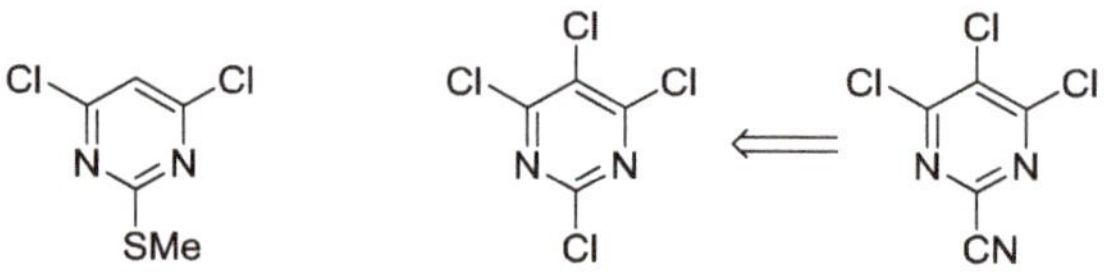

Scheme 2. Structure of 4,6-dichloro-2-(methylthio)pyrimidine (**3**) and retrosynthetic analysis of trichloropyrimidine **1**.

2. Results and Discussion

We subjected tetrachloropyrimidine **4** to a variety of displacement conditions involving the use of KCN with 18-crown-6 (0.1 equiv.), in the solvents MeCN, dioxane, DCM, or H_2O and temperature ranging between 20 and 100 °C, which led to either no reaction or degradation of the starting material. In light of this, we turned to using *n*-Bu_4NCN as the cyanide source that has been reported to afford the cyanide substitution of 4-chloropyrimidine derivatives [19]. We therefore screened this reagent in the presence of DABCO, which was used as a catalyst for the reported transformation [19]. Reaction with *n*-Bu_4NCN (2 equiv.) in the solvents MeCN, DMSO, acetone, PhH, MeOH, or even neat led to the degradation of the starting material. Similarly, biphasic systems such as DCM/H_2O or Pd-catalyzed conditions ($Pd(OAc)_2$ with the ligand dppb) [20] also led to degradation of the starting materials.

Interestingly, among our efforts to displace the C2 chloride in the presence of DABCO, we observed the formation of a colorless side-product, identified as 2,4,5-trichloro-6-[4-(2-chloroethyl) piperazin-1-yl]pyrimidine (**5**), which was isolated in 17% yield along with 60% recovered starting material (Scheme 3). Reaction of tetrachloropyrimidine **4** with 1 equiv. of DABCO in MeCN, at ca. 20 °C, gave a 42% yield of piperazine **5** as the only product (Scheme 3, see the Supplementary Materials for NMR spectra).

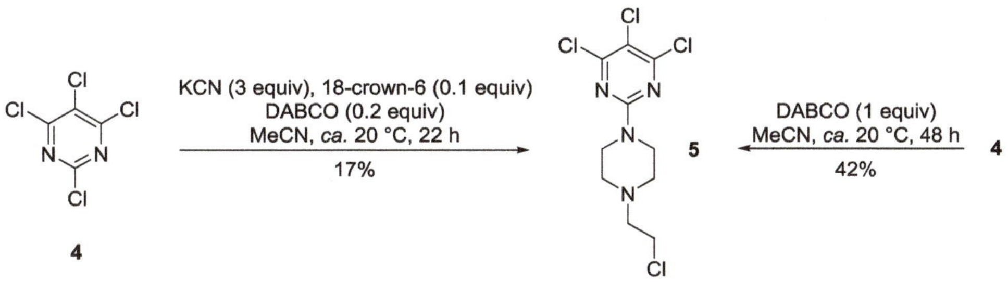

Scheme 3. Synthesis of 2,4,5-trichloro-6-[4-(2-chloroethyl)piperazin-1-yl]pyrimidine (**5**).

Intrigued by this result, we then subjected 4,5,6-trichloropyrimidine-2-carbonitrile (**1**) to the same reaction conditions that led to a slow consumption of the starting material, giving 4,5-dichloro-6-[4-(2-chloroethyl)piperazin-1-yl]pyrimidine-2-carbonitrile (**6**) as the only product in 52% yield (Scheme 4, see the Supplementary Materials for NMR spectra). The formation of the two products **5** and **6** reveals that the most reactive chloride of tetrachloropyrimidine **4**, towards DABCO, is at the C2 position, while the most reactive site in trichloropyrimidine **1** is the C4 position. This result shows that the chemistry of trichloropyrimidine **1** is complementary to other pyrimidine scaffolds and supports its potential as a synthetic scaffold.

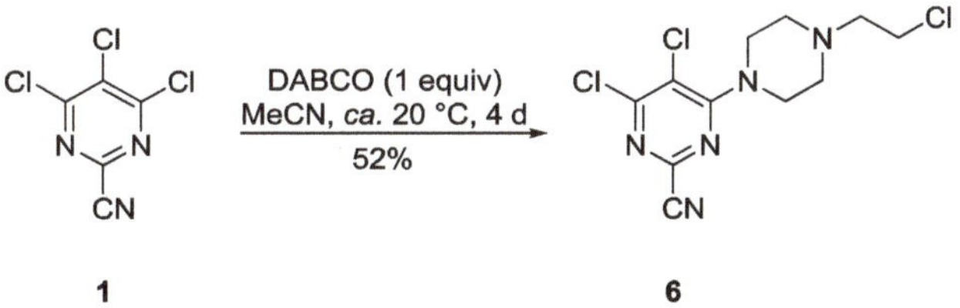

Scheme 4. Synthesis of 4,5-dichloro-6-[4-(2-chloroethyl)piperazin-1-yl]pyrimidine-2-carbonitrile (**6**).

3. Materials and Methods

The reaction mixture was monitored by TLC using commercial glass-backed thin-layer chromatography (TLC) plates (Merck Kieselgel 60 F_{254}). The plates were observed under UV light at 254 and 365 nm. Acetonitrile (MeCN) was distilled over CaH_2 before use. The melting point was determined using a PolyTherm-A, Wagner & Munz Kofler Hotstage Microscope apparatus (Wagner & Munz, Munich, Germany). The solvent used for recrystallization is indicated after the melting point. The UV-vis spectrum was obtained using a Perkin-Elmer Lambda-25 UV-vis spectrophotometer (Perkin-Elmer, Waltham, MA, USA), and inflections are identified by the abbreviation "inf". The IR spectrum was recorded on a Shimadzu FTIR-NIR Prestige-21 spectrometer (Shimadzu, Kyoto, Japan) with the Pike Miracle Ge ATR accessory (Pike Miracle, Madison, WI, USA), and strong, medium, and weak peaks are represented by s, m, and w, respectively. ^{1}H and ^{13}C NMR spectra were recorded on a Bruker Avance 500 machine (at 500 and 125 MHz, respectively, (Bruker, Billerica, MA, USA)). Deuterated solvents were used for homonuclear lock, and the signals were referenced to the deuterated solvent peaks. Attached proton test (APT) NMR studies were used for the assignment of the ^{13}C peaks as CH_3, CH_2, CH, and Cq (quarternary). The MALDI-TOF mass spectrum (+ve mode) was recorded on a Bruker Autoflex III Smartbeam instrument (Bruker). The elemental analysis was run by the London Metropolitan University Elemental Analysis Service. 4,5,6-Trichloropyrimidine-2-carbonitrile (**1**) and 2,4,5,6-tetrachloropyrimidine (**4**) were prepared according to the literature procedures [12,18].

4,5,6-Trichloro-2-[4-(2-chloroethyl)piperazin-1-yl]pyrimidine (**5**)

One portion 1,4-diazabicyclo[2.2.2]octane (DABCO, 56.0 mg, 0.500 mmol) was added to a stirred mixture of 2,4,5,6-tetrachloropyrimidine (**4**) (109 mg, 0.500 mmol) in MeCN (5 mL) at ca. 20 °C. The mixture was protected with a $CaCl_2$ drying tube and stirred at this temperature until complete consumption of the starting material (TLC, 48 h). DCM (10 mL) was then added, the mixture adsorbed onto silica, and chromatography (DCM) gave the *title compound* **5** (63.3 mg, 42%) as colorless plates, mp 84–85 °C (from MeCN); R_f 0.21 (DCM); (found: C, 36.47; H, 3.76; N, 16.86. $C_{10}H_{12}Cl_4N_4$ requires C, 36.39; H, 3.67; N, 16.98%); λ_{max}(DCM)/nm 262 (log ε 4.76), 332 (3.77); v_{max}/cm^{-1} 2955 w, 2857 w and 2810 w (C-H), 1566 s, 1520 w, 1483 m, 1450 w, 1366 w, 1302 m, 1283 m, 1196 m, 1179 w, 1144 w, 1076 w, 1001 m, 986 m, 812 m, 762 m; δ_H(500 MHz; $CDCl_3$) 3.81 (4H, *t*, *J* 5.0, pip. NCH_2), 3.61 (2H, *t*, *J* 6.8, CH_2Cl), 2.77 (2H, *t*, *J* 6.8, NCH_2), 2.56 (4H, *t*, *J* 4.8, pip. NCH_2); δ_C(125 MHz; $CDCl_3$) 159.2 (Cq), 157.2 (Cq), 113.1 (Cq), 59.6 (CH_3), 52.7 (CH_3), 44.0 (CH_3), 40.8 (CH_3); *m/z* (MALDI-TOF) 331 (MH$^+$ + 2, 80%), 329 (MH$^+$, 100), 266 (36).

4,5-Dichloro-6-[4-(2-chloroethyl)piperazin-1-yl]pyrimidine-2-carbonitrile (**6**)

One portion 1,4-diazabicyclo[2.2.2]octane (DABCO, 56.0 mg, 0.500 mmol) was added to a stirred mixture of 4,5,6-trichloropyrimidine-2-carbonitrile (**1**) (104 mg, 0.500 mmol) in MeCN (5 mL) at ca. 20 °C. The mixture was protected with a $CaCl_2$ drying tube and stirred at this temperature until complete consumption of the starting material (TLC, 4 days). DCM (10 mL) was then added, the mixture adsorbed onto silica, and chromatography (DCM/Et$_2$O, 95:5) gave the *title compound* **6** (83.4 mg, 52%) as colorless needles, mp 47–48 °C (from MeOH/−60 °C); R_f 0.73 (DCM/Et$_2$O, 95:5); (found: C, 41.27; H, 3.83; N, 21.65. $C_{11}H_{12}Cl_3N_5$ requires C, 41.21; H, 3.77; N, 21.84%); λ_{max}(DCM)/nm 237 (log ε 4.27), 285 (4.20); v_{max}/cm^{-1} 2924 w and 2853 w (C-H), 1647 s, 1468 m, 1450 m, 1445 m, 1371 m, 1302 m, 1269 m, 1234 w, 1209 w, 1161 w, 1144 m, 1128 m, 1096 m, 1040 m, 997 s, 897 m, 800 w, 766 w; δ_H(500 MHz; $CDCl_3$) 3.86 (4H, *t*, *J* 4.9, pip. NCH_2), 3.61 (2H, *t*, *J* 6.7, CH_2Cl), 2.79 (2H, *t*, *J* 6.7, NCH_2), 2.65 (4H, *t*, *J* 4.9, pip. NCH_2); δ_C(125 MHz; $CDCl_3$) 160.6 (Cq), 159.7 (Cq), 139.5 (Cq), 116.3 (Cq), 114.6 (Cq), 59.3 (CH_3), 52.7 (CH_3), 47.9 (CH_3), 40.7 (CH_3); *m/z* (MALDI-TOF) 324 (MH$^+$ + 4, 35%), 322 (MH$^+$ + 2, 72), 320 (MH$^+$, 100), 217 (11).

Supplementary Materials: The following are available online: molfile, ^{1}H and ^{13}C-NMR spectra.

Author Contributions: P.A.K. and A.S.K. conceived the experiments; A.S.K. designed and performed the experiments, analyzed the data, and wrote the paper.

Funding: This research was funded by the Cyprus Research Promotion Foundation (Grants: ΣΤΡΑΤΗΙΙ/0308/06, ΝΕΚΥΡ/0308/02 ΥΓΕΙΑ/0506/19 and ΕΝΙΣΧ/0308/83).

Acknowledgments: The authors thank the following organizations and companies in Cyprus for generous donations of chemicals and glassware: the State General Laboratory, the Agricultural Research Institute, the Ministry of Agriculture, MedoChemie Ltd., Medisell Ltd. and Biotronics Ltd. Furthermore, we thank the A. G. Leventis Foundation for helping to establish the NMR facility at the University of Cyprus.

Conflicts of Interest: The authors declare no conflict of interest. The founding sponsors had no role in the design of the study; in the collection, analyses, or interpretation of data; in the writing of the manuscript, and in the decision to publish the results.

References

1. Vitaku, E.; Smith, D.T.; Njardarson, J.T. Analysis of the Structural Diversity, Substitution Patterns, and Frequency of Nitrogen Heterocycles among U.S. FDA Approved Pharmaceuticals. *J. Med. Chem.* **2014**, *57*, 10257–10274. [CrossRef] [PubMed]
2. Gettys, K.E.; Ye, Z.; Dai, M. Recent Advances in Piperazine Synthesis. *Synthesis* **2017**, *49*, 2589–2604. [CrossRef]
3. Bugaenko, D.I. 1,4-Diazabicyclo[2.2.2]octane in the synthesis of piperazine derivatives. *Chem. Heterocycl. Com.* **2017**, *53*, 1277–1279. [CrossRef]
4. Yavari, I.; Bayat, M.J.; Ghazanfarpour-Darjani, M. Synthesis of N-alkyl-N′-aryl-piperazines via copper-catalyzed C–N bond formation. *Tetrahedron Lett.* **2014**, *55*, 5595–5596. [CrossRef]

5. Dong, H.-R.; Chen, Z.-B.; Li, R.-S.; Dong, H.-S.; Xie, Z.-X. Convenient and efficient synthesis of disubstituted piperazine derivatives by catalyst-free, atom-economical and tricomponent domino reactions. *RSC Adv.* **2015**, *5*, 10768–10772. [CrossRef]

6. Wang, H.-J.; Earley, W.G.; Lewis, R.M.; Srivastava, R.R.; Zych, A.J.; Jenkins, D.M.; Fairfax, D.J. An efficient one-pot, two-step synthesis of 4-substituted 1-heteroarylpiperazines under microwave irradiation conditions. *Tetrahedron Lett.* **2007**, *48*, 3043–3046. [CrossRef]

7. Wu, H.-Q.; Yang, K.l.; Chen, X.-Y.; Arulkumar, M.; Wang, N.; Chena, S.-H.; Wang, Z.-Y. A 3,4-dihalo-2(5*H*)-furanone initiated ringopening reaction of DABCO in the absence of a metal catalyst and additive and its application in a one-pot two-step reaction. *Green Chem.* **2019**, *21*, 3782–3788. [CrossRef]

8. Wang, H.-J.; Wang, Y.; Csakai, A.J.; Earley, W.G.; Herr, R.J. Efficient N-Arylation/Dealkylation of Electron Deficient Heteroaryl Chlorides and Bicyclic Tertiary Amines under Microwave Irradiation. *J. Comb. Chem.* **2009**, *11*, 355–363. [CrossRef] [PubMed]

9. Fu, Y.; Xu, Q.-S.; Li, Q.-Z.; Li, M.-P.; Shi, C.-Z.; Du, Z. Sulfonylation of 1,4-Diazabicyclo[2.2.2]octane: Charge-Transfer Complex Triggered C–N Bond Cleavage. *ChemistryOpen* **2019**, *8*, 127–131. [CrossRef] [PubMed]

10. Kolesinska, B.; Barszcz, K.; Kaminski, Z.J.; Drozdowska, D.; Wietrzyk, J.; Switalska, M. Synthesis and cytotoxicity studies of bifunctional hybrids of nitrogen mustards with potential enzymes inhibitors based on melamine framework. *J. Enzyme Inhib. Med.* **2011**, *27*, 619–627. [CrossRef] [PubMed]

11. Koyioni, M.; Manoli, M.; Koutentis, P.A. The Reaction of DABCO with 4-Chloro-5*H*-1,2,3-dithiazoles: Synthesis and Chemistry of 4-[*N*-(2-Chloroethyl)piperazin-1-yl]-5*H*-1,2,3-dithiazoles. *J. Org. Chem.* **2016**, *81*, 615–631. [CrossRef] [PubMed]

12. Kalogirou, A.S.; Manoli, M.; Koutentis, P.A. Reaction of 3,4,4,5-tetrachloro-4*H*-1,2,6-thiadiazine with benzyltriethylammonium chloride. *Tetrahedron Lett.* **2018**, *59*, 3589–3593. [CrossRef]

13. Kalogirou, A.S.; Koutentis, P.A. A qualitative comparison of the reactivities of 3,4,4,5-tetrachloro-4*H*-1,2,6-thiadiazine and 4,5-dichloro-1,2,3-dithiazolium chloride. *Molecules* **2015**, *20*, 14576–14594. [CrossRef] [PubMed]

14. Koutentis, P.A.; Rees, C.W.; White, A.J.P.; Williams, D.J. Reaction of tetracyanoethylene with SCl_2; new molecular rearrangements. *Chem. Commun.* **2000**, 303–304. [CrossRef]

15. Koutentis, P.A.; Rees, C.W. Reaction of tetracyanoethylene with SCl_2; new molecular rearrangements. *J. Chem. Soc., Perkin Trans. 1* **2000**, 1089–1094. [CrossRef]

16. Kalogirou, A.S.; Koutentis, P.K. 4-Chloro-6-ethoxy-2-(methylthio)pyrimidine. *Molbank* **2017**, *2017*, M923. [CrossRef]

17. Kalogirou, A.S.; Koutentis, P.K. Synthesis of 2-cyanopyrimidines. *Molbank* **2019**. submitted.

18. Childress, S.J.; McKee, R.L. Chloroaminopyrimidines. *J. Am. Chem. Soc.* **1950**, *729*, 4271–4272. [CrossRef]

19. Griffith, D.A. Purine Compounds and Uses Thereof as Cannabinoid Receptor Ligands. WO2004/37823, 6 May 2004.

20. Schareina, T.; Zapf, A.; Cotté, A.; Müller, N.; Beller, M. A Bio-inspired Copper Catalyst System for Practical Catalytic Cyanation of Aryl Bromides. *Synthesis* **2008**, *20*, 3351–3355. [CrossRef]

Communication

Synthesis of 2-Cyanopyrimidines

Andreas S. Kalogirou [1],* and Panayiotis A. Koutentis [2]

[1] Department of Life Sciences, School of Sciences, European University Cyprus, 6 Diogenis Str., Engomi, P.O. Box 22006, 1516 Nicosia, Cyprus
[2] Department of Chemistry, University of Cyprus, P.O. Box 20537, 1678 Nicosia, Cyprus; koutenti@ucy.ac.cy
*   Correspondence: A.Kalogirou@euc.ac.cy; Tel.: +357-22892804

Received: 26 September 2019; Accepted: 21 October 2019; Published: 22 October 2019

Abstract: 4,6-Dichloro-2-(methylthio)pyrimidine (**7**) was converted to 4-chloro-6-methoxy-2-(methylthio)pyrimidine (**15**) and 4,6-dimethoxy-2-(methylthio)pyrimidine (**14**). Chlorination of the latter with *N*-chlorosuccinimide (NCS) affords 5-chloro-4,6-dimethoxy-2-(methylthio)pyrimidine (**16**) in 56% yield.   Both methylthiopyrimidines **15** and **14** were converted in two steps to 4-chloro-6-methoxypyrimidine-2-carbonitrile (**13**) and 4,6-dimethoxypyrimidine-2-carbonitrile (**12**), respectively, after oxidation to sulfones and displacement of the sulfinate group with KCN. 4,6-Dimethoxypyrimidine-2-carbonitrile (**12**) was chlorinated with NCS to give 5-chloro-4,6-dimethoxypyrimidine-2-carbonitrile (**10**) in 53% yield.   All new compounds were fully characterized.

Keywords: heterocycle; pyrimidine; nucleophilic displacement; chlorination

---

1. Introduction

Pyrimidines are important aromatic N-heterocycles that are found in nature, for example, as components of pyrimidine nucleotides and vitamin B1 (thiamine).   Not surprisingly, the chemistry of pyrimidines has been investigated for over a century and numerous reviews have appeared [1]. Pyrimidines are also present in many drugs such as the CNS depressant phenobarbital, the anti-HIV agent zidovudine and the hyperthyroidism drug propylthiouracil (Figure 1). Additional pharmaceutical applications include uses as diuretics [2], anti-inflammatory [3], anti-malarial [4], and anti-tumor [5] agents.

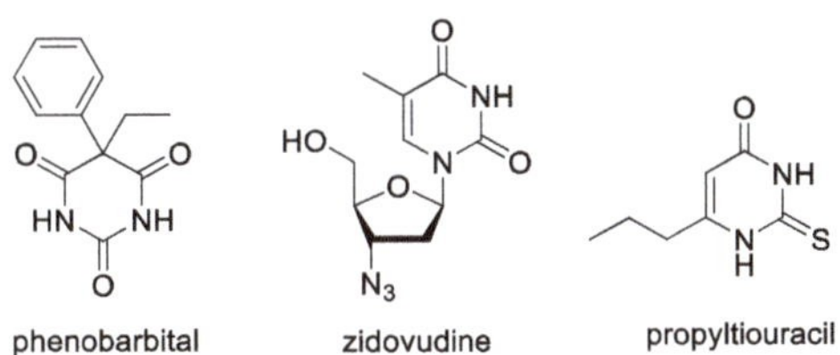

Figure 1. Pyrimidine containing drugs.

Our interest in pyrimidines began with 4,5,6-trichloropyrimidine-2-carbonitrile (**1**), which was isolated as an unexpected minor product (1–5%) from the reaction of tetracyanoethene (TCNE) with SCl$_2$ during the preparation of 2-(3,5-dichloro-4*H*-1,2,6-thiadiazin-4-ylidene)malononitrile (**2**) [6] (Scheme 1). To obtain access to larger quantities of trichloropyrimidine **1**, so that its chemistry could be investigated, we pursued various independent syntheses.

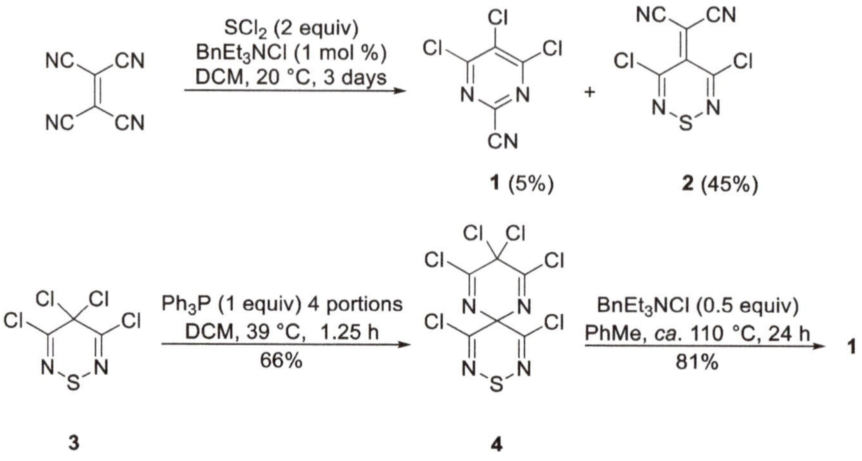

Scheme 1. Preparation of trichloropyrimidine **1** from TCNE and from tetrachlorothiadiazine **3** [6,7].

To date, our most efficient synthesis of pyrimidine **1** starts from the highly reactive tetrachlorothiadiazine **3** via perchloro-9-thia-1,5,8,10-tetraazaspiro[5.5]undeca-1,4,7,10-tetraene (**4**) in a 53% overall yield [7] (Scheme 1).

Below we report an alternative effort to prepare pyrimidine **1** that failed but did lead to the preparation of several new poly-substituted pyrimidines.

A retrosynthetic analysis of trichloropyrimidine **1** revealed that the cyano-group could be introduced by a nucleophilic displacement of a suitable leaving group in the C2 position by cyanide. Such a group could be sulfinate **5** that can be prepared by oxidation of thioether **6**. The latter could be formed by C5 chlorination of the readily available starting pyrimidine **7** (Scheme 2).

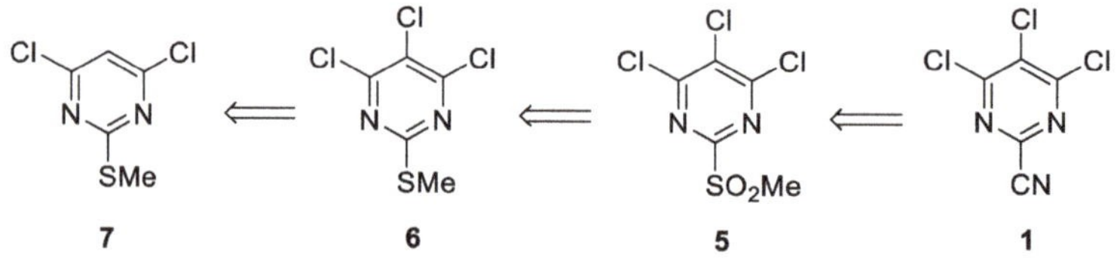

Scheme 2. Retrosynthetic analysis of trichloropyrimidine **1**.

In two alternative retrosynthetic strategies, pyrimidine **1** could be formed by 4,6-dihydroxy-pyrimidine **8** or 4-hydroxypyrimidine **9** (Scheme 3). These compounds could be formed by deprotection of the respective ethers **10** and **11**. The presence of alkoxy groups in the C5 and C6 positions would enable the easier chlorination of the C5 in precursors **12** and **13** due to the electron-donating character of these groups. The cyano groups could be introduced in a similar manner to that described above from thioethers **14** and **15**, which could be formed from chloride displacement of dichloropyrimidine **7**.

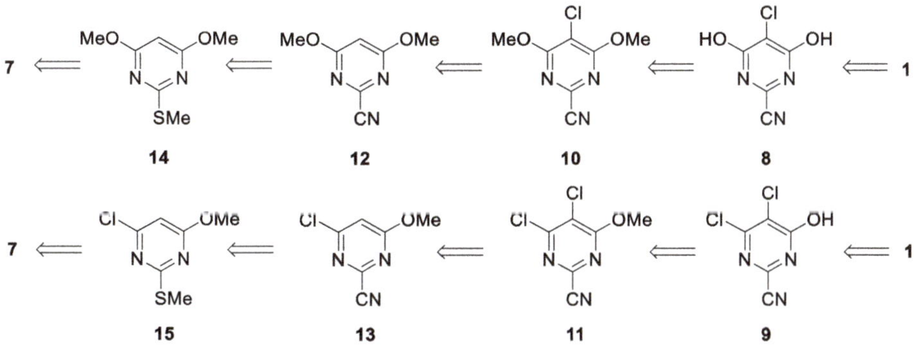

Scheme 3. Alternative retrosynthetic analysis via methoxypyrimidines 14 and 15.

2. Results and Discussion

As described above, our independent synthesis began from the known 4,6-dichloro-2-(methylthio) pyrimidine (7) that was prepared in two steps and 92% overall yield from thiobarbituric acid [8]. This starting material was selected because of the availability of the starting thiobarbituric acid and the high yield of its transformation to dichloropyrimidine 7. The latter has a versatile thioether group at C2 that can be displaced by cyanide and chlorides at the C4,6 positions. Early efforts to chlorinate the C5 position of dichloropyrimidine 7 with either NCS in AcOH, at ca. 117 °C, refluxing $PCl_5/POCl_3$ or neat PCl_5 in a sealed tube at ca. 130 °C failed, giving only recovered starting material. Tentatively, this was attributed to the electron-deficient nature of the ring in the presence of the C4/6 electronegative chlorine atoms.

As such, we 'activated' the pyrimidine C5 position towards electrophilic chlorination by displacing one or both chlorides by the strong electron-releasing alkoxides. The substitution reaction is known [9,10], but by slightly modifying the reaction procedure to involve a more concentrated reaction mixture we reduced the literature reaction time from 18 h to 2 h and obtained a high yield of methoxypyrimidine 15 (Scheme 4). Similarly, for the preparation of dimethoxy-pyrimidine 14, elevating the reaction temperature to ca. 65 °C from 20 °C also led to shorter reaction time (2 h vs. 18 h) and a high yield of 14 (Scheme 4).

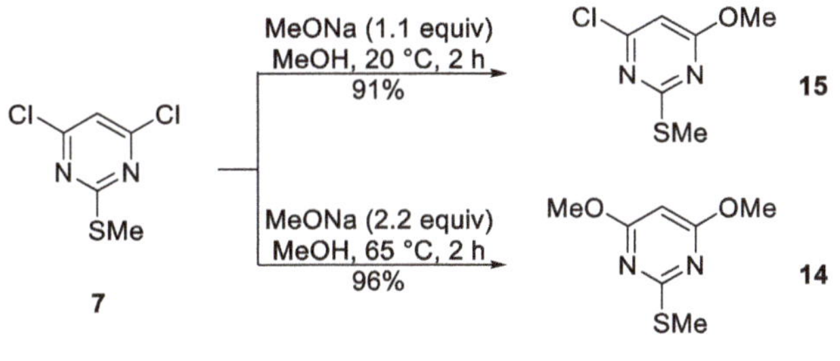

Scheme 4. Preparation of methoxypyrimidines 14 and 15.

Attempted C5 chlorination of the methoxypyrimidine 15 using either NCS or PCl_5 as chlorinating agents failed to give a complex mixture of products, but fortunately, dimethoxy-pyrimidine 14 reacted smoothly with NCS in AcOH to give the new 5-chloropyrimidine 16 in a moderate 56% yield (Scheme 5, see the supplementary materials for NMR spectra).

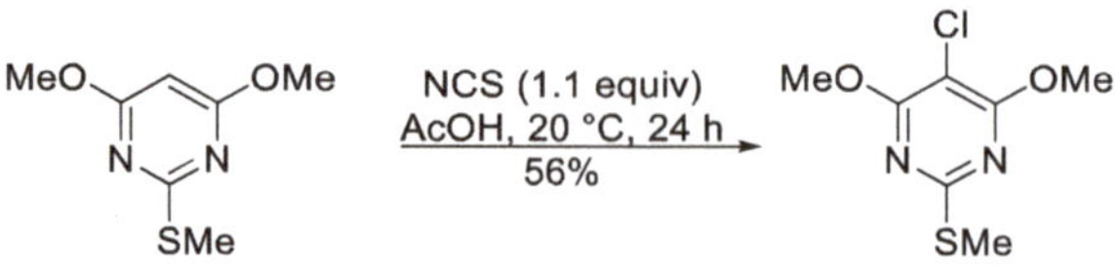

Scheme 5. Preparation of 5-chloro-4,6-dimethoxy-2-(methylthio)pyrimidine (**16**).

Encouraged by this result, we then proceeded to functionalize the C2 ring position. Oxidation of the thioether moiety in both pyrimidines **15** and **14** was performed by MCPBA (2 equiv), in DCM, at ca. 0 °C, to give sulfones **17** and **18**, respectively in excellent yields (Scheme 6). For comparison, the literature procedure for the preparation of methoxypyrimidine **17** used the oxidant oxone and obtained 78% yield of product **17** [11], whereas for dimethoxypyrimidine **18**, MCPBA was used at a temperature of ca. 30 °C giving a 67% yield of product **18** [12]. Subsequent displacement of the sulfinate with KCN in MeCN yielded the new 2-cyanopyrimidines **13** and **12** in 27 and 83% yields, respectively (Scheme 6, see the supplementary materials for NMR spectra).

Scheme 6. Preparation of 4-chloro-6-methoxypyrimidine-2-carbonitrile (**13**) and 4,6-dimethoxypyrimidine-2-carbonitrile (**12**).

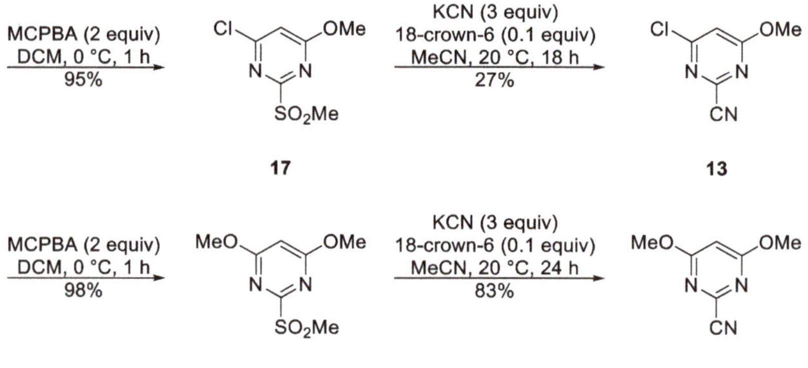

Unfortunately, attempts to chlorinate cyanopyrimidine **13** failed but the chlorination of dimethoxypyrimidine **12** was successful and gave the new 5-chloropyrimidine **10** a potential precursor to trichloropyrimidine **1** (Scheme 7, see SI for NMR spectra). Disappointingly, the subsequent step of demethylation required to reach the target compound failed. In more detail, the reaction of chloropyrimidine **10** with BBr$_3$ (5 equiv) in DCM at ca. 20 °C gave a complex mixture of products, while the use of TMSI in MeCN at ca. 82 °C led to degradation of the starting material to give tentatively acyclic side-products.

Although this study has not yielded the desired trichloropyrimidine **1**, it has given access to four new polyfunctionalized pyrimidines that could be of use for the further investigation of the chemistry and properties of pyrimidines.

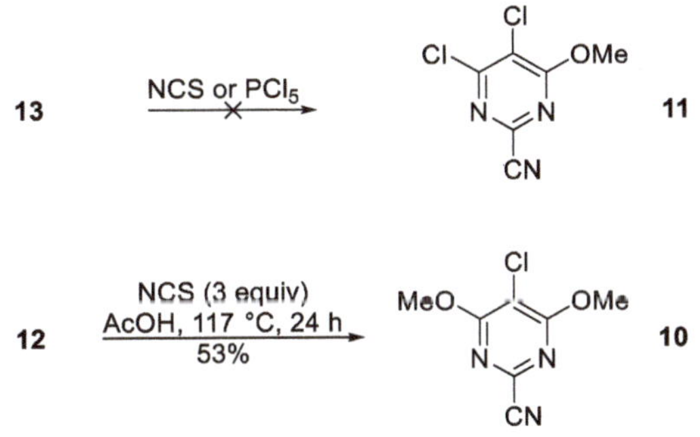

$$\textbf{13} \quad \xrightarrow[\times]{\text{NCS or PCl}_5} \quad \textbf{11}$$

$$\textbf{12} \quad \xrightarrow[53\%]{\substack{\text{NCS (3 equiv)}\\ \text{AcOH, 117 °C, 24 h}}} \quad \textbf{10}$$

Scheme 7. Preparation of 5-chloro-4,6-dimethoxypyrimidine-2-carbonitrile (**10**).

3. Materials and Methods

The reaction mixture was monitored by TLC using commercial glass-backed thin-layer chromatography (TLC) plates (Merck Kieselgel 60 F_{254}). The plates were observed under UV light at 254 and 365 nm. Dichloromethane (DCM) and acetonitrile (MeCN) were distilled over CaH_2 before use. The 1 M and 3 M solutions of sodium methoxide (MeONa) in methanol (MeOH) were freshly prepared by the reaction of sodium metal with MeOH. The melting point was determined using a PolyTherm-A, Wagner & Munz, Kofler Hotstage Microscope apparatus (Wagner & Munz, Munich, Germany). The solvent used for recrystallization is indicated after the melting point. The UV-vis spectrum was obtained using a Perkin-Elmer Lambda-25 UV-vis spectrophotometer (Perkin-Elmer, Waltham, MA, USA), and inflections are identified by the abbreviation "inf". The IR spectrum was recorded on a Shimadzu FTIR-NIR Prestige-21 spectrometer (Shimadzu, Kyoto, Japan) with Pike Miracle Ge ATR accessory (Pike Miracle, Madison, WI, USA) and strong, medium and weak peaks are represented by s, m and w, respectively. ^{1}H and ^{13}C NMR spectra were recorded on a Bruker Avance 500 spectrometer [at 500 and 125 MHz, respectively, (Bruker, Billerica, MA, USA)]. Deuterated solvents were used for homonuclear lock and the signals are referenced to the deuterated solvent peaks. Attached proton test (APT) NMR studies were used for the assignment of the ^{13}C peaks as CH_3, CH_2, CH, and Cq (quaternary). The MALDI-TOF mass spectrum (+ve mode) was recorded on a Bruker Autoflex III Smartbeam instrument (Bruker). The elemental analysis was run by the London Metropolitan University Elemental Analysis Service. 4,6-dichloro-2-(methylthio)pyrimidine (**7**) was prepared according to the literature procedure [8].

4-Chloro-6-methoxy-2-(methylthio)pyrimidine (**15**)

To a stirred mixture of 4,6-dichloro-2-(methylthio)pyrimidine (**7**) (585 mg, 3.00 mmol) in MeOH (15 mL) at ca. 20 °C was added in one portion a solution of MeONa 1 M in MeOH (3.30 mL, 3.30 mmol). The mixture was protected with a $CaCl_2$ drying tube and stirred at this temperature until complete consumption of the starting material (TLC, 2 h). Et_2O (20 mL) and $NaHCO_3$ sat. (10 mL) were then added, the two layers separated, and the aqueous layer extracted with a further 10 mL of Et_2O. The combined organic phases were then dried over Na_2SO_4, filtered, and evaporated in vacuo to give the title compound **15** (518 mg, 91%) as a colorless oil; R_f 0.31 (*n*-hexane/DCM, 70:30), δ_H(500 MHz; $CDCl_3$) 6.41 (1H, s, C*H*), 3.96 (3H, s, OC*H₃*), 2.55 (3H, s, SC*H₃*), identical to the one reported [11].

4,6-Dimethoxy-2-(methylthio)pyrimidine (**14**)

To a stirred mixture of 4,6-dichloro-2-(methylthio)pyrimidine (**7**) (780 mg, 4.00 mmol) in MeOH (15 mL) at ca. 20 °C was added in one portion a solution of MeONa 3 M in MeOH (2.93 mL, 8.80 mmol). The mixture was protected with a $CaCl_2$ drying tube and heated to ca. 65 °C until complete consumption of the starting material (TLC, 2 h). DCM (10 mL) was then added, the mixture adsorbed onto silica, and chromatography (*n*-hexane/DCM 70:30) gave the title compound **14** (717 mg, 96%) as yellow plates, mp 53–54 °C (from *n*-hexane/−40 °C, lit. 53–54 °C [13]); R$_f$ 0.18 (*n*-hexane/DCM, 70:30); δ_H(500 MHz; CDCl$_3$) 5.71 (1H, s, C*H*), 3.92 (6H, s, OC*H$_3$*), 2.54 (6H, s, SC*H$_3$*), identical to the one reported [14].

5-Chloro-4,6-dimethoxy-2-(methylthio)pyrimidine (**16**)

To a stirred mixture of 4,6-dimethoxy-2-(methylthio)pyrimidine (**14**) (93.1 mg, 0.500 mmol) in AcOH (1 mL) at ca. 20 °C was added in one portion *N*-chlorosuccinimide (73.4 mg, 0.55 mmol). The mixture was protected with a CaCl$_2$ drying tube and stirred at this temperature until complete consumption of the starting material (TLC, 24 h). DCM (10 mL) was then added, the mixture adsorbed onto silica, and chromatography (*n*-hexane/DCM 70:30) gave the *title compound* **16** (61.6 mg, 56%) as colorless needles, mp 128–129 °C (from *n*-hexane/−40 °C); R$_f$ 0.46 (*n*-hexane/DCM, 70:30); (found: C, 37.94; H, 4.12; N, 12.53. C$_7$H$_9$ClN$_2$O$_2$S requires C, 38.10; H, 4.11; N, 12.69%); λ_{max}(DCM)/nm 258 inf (log ε 4.11), 267 (4.16); v_{max}/cm^{-1} 2963w and 2930w (C–H), 1560m, 1555s, 1493m, 1458w, 1389m, 1358s, 1335w, 1319m, 1290m, 1269m, 1186m, 1182m, 1126s, 1070m, 926m, 770m; δ_H(500 MHz; CDCl$_3$) 4.03 (6H, s, OC*H$_3$*), 2.53 (3H, s, SC*H$_3$*); δ_C(125 MHz; CDCl$_3$) 167.5 (*Cq*), 165.1 (*Cq*), 95.2 (*Cq*), 55.0 (*CH$_3$*), 14.4 (*CH$_3$*); *m/z* (MALDI-TOF) 221 (M$^+$-H + 2, 33%), 220 (M$^+$, 75), 219 (M$^+$ − H, 100), 206 (12).

4-Chloro-6-methoxy-2-(methylsulfonyl)pyrimidine (**17**)

To a stirred mixture of 4-chloro-6-methoxy-2-(methylthio)pyrimidine (**15**) (572 mg, 3.00 mmol) in DCM (10 mL) cooled in an ice-bath to ca. 0 °C was added in one portion m-chloroperbenzoic acid of 77% purity (1.344 g, 6.000 mmol). The mixture was protected with a CaCl$_2$ drying tube and stirred at this temperature until complete consumption of the starting material (TLC, 1 h). Et$_2$O (20 mL) and Na$_2$CO$_3$ sat. (10 mL) were then added, the two layers separated, and the aqueous layer extracted with a further 10 mL of Et$_2$O. The combined organic phases were then dried over Na$_2$SO$_4$, filtered, and evaporated in vacuo to give the title compound **17** (634 mg, 95%) as colorless needles, mp 85–86 °C (from *n*-hexane/−40 °C, lit. 87–88 °C [11]); R$_f$ 0.22 (*n*-hexane/DCM, 20:80); δ_H(500 MHz; CDCl$_3$) 6.92 (1H, s, C*H*), 4.10 (3H, s, OC*H$_3$*), 3.33 (3H, s, SO$_2$C*H$_3$*); identical to the one reported [11].

4,6-Dimethoxy-2-(methylsulfonyl)pyrimidine (**18**)

To a stirred mixture of 4,6-dimethoxy-2-(methylthio)pyrimidine (**14**) (186 mg, 1.00 mmol) in DCM (5 mL) cooled in an ice-bath to ca. 0 °C was added in one portion m-chloroperbenzoic acid of 77% purity (448 g, 3.000 mmol). The mixture was protected with a CaCl$_2$ drying tube and stirred at this temperature until complete consumption of the starting material (TLC, 1 h). Et$_2$O (20 mL) and Na$_2$CO$_3$ sat. (10 mL) were then added, the two layers separated, and the aqueous layer extracted with a further 10 mL of Et$_2$O. The combined organic phases were then dried over Na$_2$SO$_4$, filtered and evaporated in vacuo to give the title compound **18** (124 mg, 98%) as colorless plates, mp 125–126 °C (from EtOH, lit. 126–128 °C [13]); R$_f$ 0.74 (DCM); δ_H(500 MHz; CDCl$_3$) 6.18 (1H, s, C*H*), 4.03 (6H, s, OC*H$_3$*), 3.32 (3H, s, SO$_2$C*H$_3$*); identical to the one reported [13].

4-Chloro-6-methoxypyrimidine-2-carbonitrile (**13**)

To a stirred mixture of 4-chloro-6-methoxy-2-(methylsulfonyl)pyrimidine (**17**) (223 mg, 1.00 mmol) in MeCN (5 mL) at ca. 20 °C was added in one portion 18-crown-6 (26 mg, 0.10 mmol) followed by KCN (195 mg, 3.00 mmol). The mixture was protected with a CaCl$_2$ drying tube and stirred at this temperature until complete consumption of the starting material (TLC, 18 h). Et$_2$O (20 mL) and H$_2$O (10 mL) were then added, the two layers separated, and the aqueous layer extracted with a further 10 mL of Et$_2$O. The combined organic phases were then dried over Na$_2$SO$_4$, filtered, and the mixture adsorbed onto silica and chromatography (*n*-hexane/DCM 20:80) gave the *title compound* **13** (48 mg, 27%) as colorless needles, mp 69–70 °C (from *n*-hexane/−40 °C); R$_f$ 0.67 (*n*-hexane/DCM 20:80); (found: C, 42.65; H, 2.40; N, 24.63. C$_6$H$_4$ClN$_3$O requires C, 42.50; H, 2.38; N, 24.78%); λ_{max}(DCM)/nm 241 (log ε 3.59), 262 (3.41); v_{max}/cm^{-1} 3084w (C–H), 1562s, 1533m, 1518m, 1470m, 1396m, 1360s, 1341m, 1244w, 1190m, 1130s, 1038s, 978s, 945m, 880m, 862m, 770w; δ_H(500 MHz; CDCl$_3$) 6.95 (1H, s, C*H*), 4.06 (3H, s,

OCH_3); δ_C(125 MHz; CDCl$_3$) 170.8 (Cq), 161.3 (Cq), 143.5 (Cq), 114.6 (Cq), 111.1 (CH), 55.7 (CH$_3$); *m/z* (MALDI-TOF) 170 (M$^+$-H + 2, 20%), 169 (M$^+$, 40), 168 (M$^+$ − H, 100), 132 (36).

4,6-Dimethoxypyrimidine-2-carbonitrile (**12**)

To a stirred mixture of 4,6-dimethoxy-2-(methylsulfonyl)pyrimidine (**18**) (218 mg, 1.00 mmol) in MeCN (5 mL) at ca. 20 °C was added in one portion 18-crown-6 (26 mg, 0.10 mmol) followed by KCN (195 mg, 3.00 mmol). The mixture was protected with a CaCl$_2$ drying tube and stirred at this temperature until complete consumption of the starting material (TLC, 24 h). Et$_2$O (20 mL) and H$_2$O (10 mL) were then added, the two layers separated, and the aqueous layer extracted with a further 10 mL of Et$_2$O. The combined organic phases were then dried over Na$_2$SO$_4$, filtered, and the mixture adsorbed onto silica and chromatography (*n*-hexane/DCM 60:40) gave the *title compound* **12** (138 mg, 83%) as colorless needles, mp 113–114 °C (from *n*-hexane/−40 °C); R$_f$ 0.31 (*n*-hexane/DCM, 60:40); (found: C, 50.77; H, 4.35; N, 25.26. C$_7$H$_7$N$_3$O$_2$ requires C, 50.91; H, 4.27; N, 25.44%); λ_{max}(DCM)/nm 259 (log ε 3.69); v_{max}/cm^{-1} 3094w, 3003w, 2926w and 2849w (C-H), 1591s, 1566m, 1530m, 1468m, 1387m, 1352m, 1202s, 1173m, 1059s, 984m, 964m, 864m, 772m; δ_H(500 MHz; CDCl$_3$) 6.20 (1H, s, C*H*), 3.98 (6H, s, OC*H$_3$*); δ_C(125 MHz; CDCl$_3$) 171.2 (Cq), 142.6 (Cq), 115.5 (Cq), 94.1 (CH), 54.9 (CH$_3$); *m/z* (MALDI-TOF) 166 (MH$^+$, 100%), 165 (M$^+$, 25), 164 (M$^+$ − H, 51), 162 (55).

5-Chloro-4,6-dimethoxypyrimidine-2-carbonitrile (**10**)

To a stirred mixture of 4,6-dimethoxypyrimidine-2-carbonitrile (**12**) (50.0 mg, 0.303 mmol) in AcOH (2 mL) at ca. 20 °C was added in one portion *N*-chlorosuccinimide (121 mg, 0.908 mmol). The mixture was protected with a CaCl$_2$ drying tube and stirred at ca. 117 °C until complete consumption of the starting material (TLC, 24 h). Et$_2$O (20 mL) and H$_2$O (10 mL) were then added, the two layers separated, and the aqueous layer extracted with a further 10 mL of Et$_2$O. The combined organic phases were then dried over Na$_2$SO$_4$, filtered, and the mixture adsorbed onto silica and chromatography (*n*-hexane/DCM 60:40) gave the *title compound* **10** (31.8 mg, 53%) as colorless needles, mp 145–147 °C (from *n*-hexane/−40 °C); R$_f$ 0.48 (*n*-hexane/DCM, 60:40); (found: C, 41.96; H, 2.89; N, 20.96. C$_7$H$_6$ClN$_3$O$_2$ requires C, 42.12; H, 3.03; N, 21.05%); λ_{max}(DCM)/nm 278 (log ε 3.66), 290 (3.45); v_{max}/cm^{-1} 3003w, 2980w and 2918w (C-H), 1572s, 1566s, 1545w, 1524w, 1493w, 1462m, 1443w, 1385m, 1373s, 1362m, 1294m, 1182m, 1126s, 1088w, 1076w, 1013w, 968m, 914m, 781s, 741w, 712m; δ_H(500 MHz; CDCl$_3$) 4.09 (6H, s, OCH$_3$); δ_C(125 MHz; CDCl$_3$) 166.0 (Cq), 138.5 (Cq), 115.1 (Cq), 104.4 (Cq), 56.1 (CH$_3$); *m/z* (MALDI-TOF) 202 (MH$^+$ + 2, 34%), 201 (M$^+$ + 2,29), 200 (M$^+$,38), 199 (M$^+$, 100), 192 (41), 156 (10).

Supplementary Materials: The following are available online, ^{1}H and ^{13}C NMR spectra.

Author Contributions: P.A.K. and A.S.K. conceived the experiments; A.S.K. designed and performed the experiments, analyzed the data and wrote the paper.

Funding: The research was funded by the Cyprus Research Promotion Foundation (Grants: ΣΤΡΑΤΗΙΙ/0308/06, ΝΕΚΥΡ/0308/02 ΥΓΕΙΑ/0506/19 and ΕΝΙΣΧ/0308/83).

Acknowledgments: The authors thank the following organizations and companies in Cyprus for generous donations of chemicals and glassware: the State General Laboratory, the Agricultural Research Institute, the Ministry of Agriculture, MedoChemie Ltd., Medisell Ltd. and Biotronics Ltd. Furthermore, we thank the A. G. Leventis Foundation for helping to establish the NMR facility at the University of Cyprus.

Conflicts of Interest: The authors declare no conflict of interest. The founding sponsors had no role in the design of the study; in the collection, analyses, or interpretation of data; in the writing of the manuscript, and in the decision to publish the results.

References

1. Brown, D.J. Pyrimidines and their Benzo Derivatives. In *Comprehensive Heterocyclic Chemistry*; Katritzky, A.R., Rees, C.W., Eds.; Pergamon Press: Oxford, UK, 1984; Volume 3, pp. 57–155.
2. Ukrainets, I.V.; Tugaibei, I.A.; Bereznykova, N.L.; Karvechenko, V.N.; Turov, A.V. 4-Hydroxy-2-quinolones 144. Alkyl-, arylalkyl-, and arylamides of 2-hydroxy-4-oxo-4*H*-pyrido[1,2-*a*]pyrimidine-3-carboxylic acid and their diuretic properties. *Chem. Heterocycl. Com.* **2008**, *44*, 565–575. [CrossRef]

3. Amr, A.E.; Nermien, M.S.; Abdulla, M.M. Synthesis, reactions, and anti-inflammatory activity of heterocyclic systems fused to a thiophene moiety using citrazinic acid as synthon. *Monatsh. Chem.* **2007**, *138*, 699–707. [CrossRef]

4. Gorlitzer, K.; Herbig, S.; Walter, R.D. Indeno[1,2-*d*]pyrimidin-4-yl-amines. *Pharmazie* **1997**, *52*, 670–672.

5. Wagner, E.; Al-Kadasi, K.; Zimecki, M.; Sawka-Dobrowolska, W. Synthesis and pharmacological screening of derivatives of isoxazolo[4,5-*d*]pyrimidine. *Eur. J. Med. Chem.* **2008**, *43*, 2498–2504. [CrossRef] [PubMed]

6. Koutentis, P.A.; Rees, C.W. Reaction of tetracyanoethylene with SCl_2; new molecular rearrangements. *J. Chem. Soc. Perkin Trans.* **2000**, 1089–1094. [CrossRef]

7. Kalogirou, A.S.; Manoli, M.; Koutentis, P.A. Two-step conversion of 3,4,4,5-tetrachloro-4H-1,2,6-thiadiazine into 4,5,6-trichloropyrimidine-2-carbonitrile. *Tetrahedron Lett.* **2017**, *58*, 2618–2621. [CrossRef]

8. Raboisson, P.; Belfrage, A.; Classon, B.; Lindquist, K.; Nilsson, K.; Rosenquist, A.; Samuelson, B.; Wahling, H. Pyrimidine Substituted Macrocyclic HCV Inhibitors. U.S. Patent WO2008/095999A1, 14 August 2008.

9. Seganish, W.M.; Fischmann, T.O.; Sherborne, B.; Matasi, J.; Lavey, B.; McElroy, W.T.; Tulshian, D.; Tata, J.; Sondey, C.; Garlisi, C.G.; et al. Discovery and Structure Enabled Synthesis of 2,6-Diaminopyrimidin-4-one IRAK4 Inhibitors. *ACS Med. Chem. Lett.* **2015**, *6*, 942–947. [CrossRef] [PubMed]

10. Dixson, J.A.; Murugesan, N.; Barnes, K.D. Herbicidal Substituted Benzoylsulfonamides. U.S. Patent US5149357A, 22 September 1992.

11. Huang, T.-H.; Zhou, S.-S.; Wu, X.; An, L.; Yin, X.-X. Convenient synthesis of 2-(methylsulfonyl)pyrimidine derivatives. *Synth. Commun.* **2018**, *48*, 714–720. [CrossRef]

12. Zhang, J.; Zhang, L.; Zhou, Y.; Guo, Y.-L. A novel pyrimidine-based stable-isotope labeling reagent and its application to quantitative analysis using matrix-assisted laser desorption/ionization mass spectrometry. *J. Mass Spectrom.* **2007**, *42*, 1514–1521. [CrossRef] [PubMed]

13. Xu, D.; Zhu, X.; Xu, H.; Wang, Z. A facile synthesis of 4,6-dimethoxy-2-methylsulfonylpyrimidine. *Asian J. Chem.* **2014**, *26*, 313–314. [CrossRef]

14. Thomann, A.; Eberhard, J.; Allegretta, G.; Empting, M.; Hartmann, R.W. Mild and Catalyst-Free Microwave-Assisted Synthesis of 4,6-Disubstituted 2-Methylthiopyrimidines–Exploiting Tetrazole as an Efficient Leaving Group. *Synlett* **2015**, *26*, 2606–2610. [CrossRef]

 *molbank*

Short Note

2,2′-((1,4-Dimethoxy-1,4-dioxobutane-2,3-diylidene) bis(azanylylidene))bis(quinoline-3-carboxylic acid)

Joanna Fedorowicz [1],*, Karol Gzella [1], Paulina Wiśniewska [2] and Jarosław Sączewski [2]

[1] Department of Chemical Technology of Drugs, Faculty of Pharmacy, Medical University of Gdańsk, Al. Gen. J. Hallera 107, Gdańsk 80-416, Poland; karol.gzella@gumed.edu.pl
[2] Department of Organic Chemistry, Faculty of Pharmacy, Medical University of Gdańsk, Al. Gen. J. Hallera 107, Gdańsk 80-416, Poland; pauwis93@gumed.edu.pl (P.W.); js@gumed.edu.pl (J.S.)
* Correspondence: jfedorowicz@gumed.edu.pl; Tel.: +48-58-349-1957

Received: 27 October 2019; Accepted: 19 November 2019; Published: 21 November 2019

Abstract: The title compound, 2,2′-((1,4-dimethoxy-1,4-dioxobutane-2,3-diylidene)bis(azanylylidene)) bis(quinoline-3-carboxylic acid) was synthesized from isoxazolo[3,4-*b*]quinolin-3(1*H*)-one and dimethyl acetylenedicarboxylate (DMAD) via a double aza-Michael addition followed by [1,3]-H shifts. The product was characterized by infrared and nuclear magnetic resonance spectroscopy, as well as elemental analysis and high-resolution mass spectrometry (HRMS). The proposed reaction mechanism was rationalized by density functional theory (DFT) calculations.

Keywords: [1,3]-H shift; aza-Michael addition; DFT calculations; dimethyl acetylenedicarboxylate; isoxazolo[3,4-*b*]quinolin-3(1*H*)-one

1. Introduction

Isoxazol-3(2*H*)-one and isoxazol-5(2*H*)-one derivatives (Figure 1) represent an extensive class of heterocyclic ring systems found in natural products and building blocks employed in medicinal chemistry. They may be treated as useful tools in organic synthesis since they are small and easy to functionalize molecules that can be utilized to design novel bioactive compounds. It has been proven that these synthetic products exhibit antibacterial [1–8], antifungal [7–16], antitubercular [3], anticancer [3,6,17,18], antileucemic [5], antinflammatory [19–21], antiviral [22], anticonvulsant [1], antioxidant [2,7], and antiandrogenic [23,24] properties. They may act as inhibitors of p38 MAP kinases [25], protein kinase C [26], and protein-tyrosine phosphatase 1B, which consequently cause antiobesity effect [27,28]. They also find applications as GABA$_A$ receptor ligands [29] and glutamate receptor agonists [30–32], therefore they can affect learning and memory processes. Despite the molecular mechanism of antifungal action of isoxazol-5(2*H*)-ones such as TAN-950A [16] and drazoxolon [8,15] has not been established, it should be underlined that structurally similar antimicrobial oxazolidones such as posizolid, tedizolid, radezolid and linezolid or cycloserine serve as inhibitors of protein synthesis that prevent binding of *N*-formylmethionyl-tRNA to the ribosome. [33].

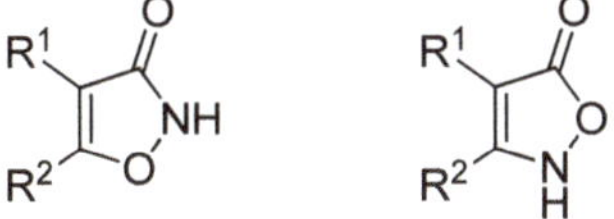

Figure 1. The structure of isoxazolone derivatives.

Recently, our research group reported antibacterial [34] and antifungal [35] *N*-substituted derivatives of 4,6-dimethylisoxazolo[3,4-*b*]pyridin-3(1*H*)-one **1**, which were obtained in N1-alkylation,

N1-acylation, and N1-sulfonation reactions. Moreover, we have implemented compound **1** and its benzo analogue, isoxazolo[3,4-*b*]quinolone-3(1*H*)-one **2**, to tandem Mannich—electrophilic amination reactions with formaldehyde and (fluoro)quinolones as secondary amines to obtain hybrid quinolone-based quaternary ammonium compounds with proved antibacterial and antibiofilm activities along with enhanced hydrophilic properties [36,37]. Furthermore, our studies aimed at reactivity exploration of isoxazolones proved that under alkaline conditions the said heterocyclic ring system (**2**) undergoes N1-alkylation followed by bimolecular base-catalyzed acyl-oxygen cleavage ($B_{AC}2$) and *O*-alkylation to yield *N,O*-dialkyl hydroxylamines (Scheme 1) [38]. Finally, we found that compounds **1** and **2** react with α,β-acetylenic carbonyl compounds (Michael acceptors) to give *N*-vinyl isoxazolones, which transform into *N*-vinylhydroxylamines by means of $B_{AC}2$ cleavage of C-O bond. The later react with an excess of Michael acceptors to yield *N,O*-divinylhydroxylamines (enamines) that undergo [3,3]-sigmatropic rearrangement to give Paal–Knorr intermediates (Scheme 1) [39]. These findings prompted us to examine the aza-Michael addition reaction of isoxazolone **2** to double activated electron-deficient acetylene, i.e., dimethyl acetylenedicarboxylate (DMAD).

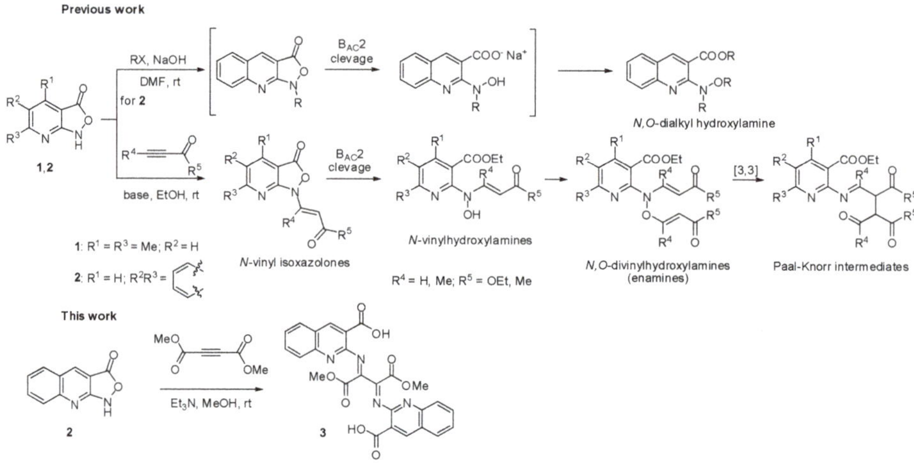

Scheme 1. Base-promoted reactivity of isoxazolones.

Herein, we describe a facile approach to 2,2′-((1,4-dimethoxy-1,4-dioxobutane-2,3-diylidene)bis (azanylylidene))bis(quinoline-3-carboxylic acid) **3** along with its characterization by experimental methods such as ^{1}H-NMR, IR, HRMS, and elemental analysis, as well as theoretical DFT calculations.

2. Results and discussion

2.1. Chemistry

We have performed the reaction of isoxazolone **2** with DMAD in anhydrous methanol at room temperature. Triethylamine was used as a base to deprotonate the acidic isoxazolone ring and hence form an ambident nucleophile capable to serve as a Michael donor. We reasoned that in the case of the reaction between compound **2** and double activated acetylene derivatives such as DMAD, dual addition to the Michael acceptor would occur (product **A**, Scheme 2). Unexpectedly, the isolated compound proved alternate structure as evidenced by ^{1}H-NMR spectrum (Supplementary Materials). The analysis of the spectrum revealed a lack of the aliphatic signal of the methine CH groups as depicted in structure **A**. Instead, two acidic protons were observed as broad singlet at $\delta = 13.85$. Presumably, the initial step of the reaction sequence involved nucleophilic attacks of isoxazolones on the α,β-unsaturated carbonyl compound. The initially formed intermediate **A** undergoes base-promoted [1,3]-hydrogen shifts with isoxazolone rings cleavage to yield 2,3-diiminosuccinate derivative **3**. The elucidated structure was supported by the results of the high-resolution mass spectrometry and elemental analysis.

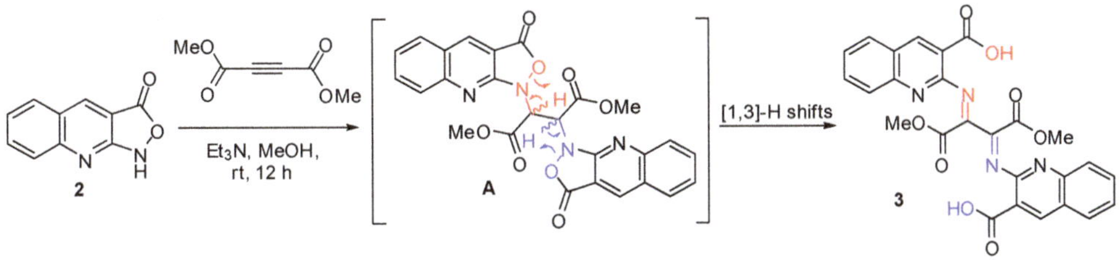

Scheme 2. The synthesis of **3** (2,2′-((1,4-dimethoxy-1,4-dioxobutane-2,3-diylidene)bis(azanylylidene)) bis(quinoline-3-carboxylic acid).

2.2. DFT Calculations

Quantum-chemical calculations were carried out to rationalize the formation of product **3**. The results obtained with density functional calculations (DFT) revealed that compound **3** is more favorable than the double Michael addition product **A**, both in gas phase and in the solvent (Table 1). Hence, the data obtained with use of B3LYP, ωB97XD, and APFD methods indicate that isomer **3** is 36.8 to 51.4 kcal/mol more stable than the intermediate **A**. Albeit the predictions are to some extent sensitive to variation of the functional type applied, the results unequivocally prove that the base-promoted [1,3] proton shifts that comprise isoxazolone ring opening transformations are intensely exothermic.

Table 1. Relative electronic energies (ΔE), and Gibbs free energies (ΔG) for isomers **A** and **3** calculated using B3LYP, ωB97XD, and APFD density functionals and 6-31G+(d) basis set in vacuum and with PCM (MeOH) solvation model.

Relative Energy	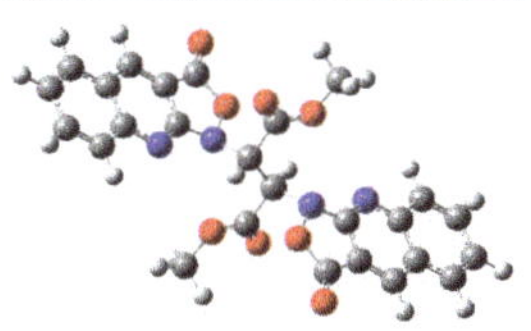	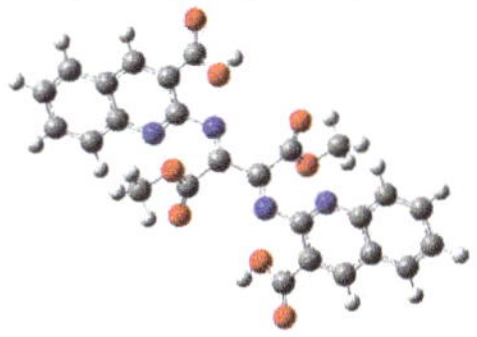
Vacuum	**A**	**3**
ΔE (kcal/mol)		
B3LYP	0	−46.5
ωB97XD	0	−40.5
APFD	0	−36.8
ΔG (kcal/mol)		
B3LYP	0	−50.9
ωB97XD	0	−45.7
APFD	0	−40.6
MeOH	**A**	**3**
ΔE (kcal/mol)		
B3LYP	0	−47.8
ωB97XD	0	−42.2
APFD	0	−38.4
ΔG (kcal/mol)		
B3LYP	0	−51.4
ωB97XD	0	−47.4
APFD	0	−41.1

3. Materials and Methods

3.1. General Methods

All reagents and solvents were purchased from commercial sources (Acros Organics, Geel, Belgium; Alfa-Aesar, Haverhill, MA, USA; or Sigma-Aldrich, Saint Louis, MO, USA) and used without further purification. Isoxazolo[3,4-*b*]quinolin-3(1*H*)-one **2** was obtained according to the known procedure [40,41]. Analytical TLC was performed on silica gel Merck 60 F254 plates (0.25 mm) with UV light visualization. Melting point was determined on an X-4 melting point apparatus with a microscope and was uncorrected. The IR spectrum was recorded on a Thermo Scientific Nicolet 380 FT-IR spectrometer. The ^{1}H NMR spectrum was registered on a Varian Unity Plus 500 MHz spectrometer. ^{1}H NMR data were internally referenced to DMSO-d_6 (2.50 ppm). The ESI-MS spectra were recorded on Shimadzu single quadrupole LCMS 2010 eV mass spectrometer (Shimadzu, Kyoto, Japan). The HRMS spectra were obtained using Agilent LC/MS Q-TOF 6550 mass spectrometer (Agilent Technologies, Santa Clara, CA, USA). Elemental analysis was performed with Elementar Vario El Cube CHNS (Elementar Analysensysteme GmbH, Langenselbold, Germany).

3.2. Synthesis of 2,2′-((1,4-dimethoxy-1,4-dioxobutane-2,3-diylidene)bis(azanylylidene))bis(quinoline-3-carboxylic acid) (**3**)

Isoxazolo[3,4-*b*]quinolin-3(1*H*)-one **2** (0.186 g, 1 mmol), dimethyl acetylenedicarboxylate (0.123 mL, 1 mmol), and triethylamine (0.139 mL, 1 mmol) were dissolved in 10 mL of anhydrous methanol. The reaction was stirred at room temperature and the reaction progress was monitored by TLC (chloroform). After 12 h, the reaction mixture was concentrated to 5 mL under reduced pressure. Upon cooling, the precipitated solid was filtered off and washed with diethyl ether (3 × 3 mL). The product was obtained as a yellow solid. Yield 0.162 g (63%); mp 240 °C (with decomposition); IR (KBr) v_{max} 2952, 1747, 1736, 1627, 1610, 1574, 1557, 1454, 1204, 1133, 1064, 787, 759 cm^{-1}; ^{1}H-NMR (500 MHz, DMSO-D$_6$) δ 3.76 (s, 6H, OCH$_3$), 7.35 (t, J = 7.8 Hz, 2H, CH), 7.41 (d, J = 7.8 Hz, 2H, CH), 7.79 (t, J = 7.8 Hz, 2H, CH), 7.99 (d, J = 7.8 Hz, 2H, CH), 8.71 (s, 2H, CH), 13.65 (bs, 2H, COOH); ESI-MS positive ionization *m/z* 515 [M + 1]$^+$, negative ionization *m/z* 513 [M − 1]$^-$; HRMS *m/z* 515.1194 [M + 1]$^+$ (calcd for C$_{26}$H$_{19}$N$_4$O$_8$$^+$, 515.1197); anal. C 59.61, H 3.59, N 10.74%, calcd for C$_{26}$H$_{18}$N$_4$O$_8$·0.5H$_2$O, C 59.66, H 3.66, N 10.70%.

3.3. DFT Calculations

All calculations have been performed with the Gaussian 16 [42] using standard algorithms and thresholds. The hybrid Becke-3–Lee–Yang–Parr functional (B3LYP) [43], long-range-corrected hybrid functional ωB97XD [44], as well as Austin–Frisch–Petersson hybrid density functional with dispersion (APFD) [45] were utilized. The bulk solvent effects were taken into account for the DFT calculations by means of polarizable continuum model (IEF-PCM) [46]. The 6-31G+(d) standard basis set has been used in the course of this study. The geometry optimizations for the studied molecules were carried out in their ground states with the inclusion of solvent effects. Vibrational analyses were used to verify that the optimized structures correspond to local minima on the energy surface. Gibbs energies including zero-point corrections, temperature corrections, and vibrational energies were computed for standard conditions (T = 298.15 K, P = 1.0 atm) using the harmonic oscillator approximation.

4. Conclusions

In summary, we have shown that isoxazolone **2** in the presence of triethylamine reacts with highly electrophilic DMAD via double aza-Michael addition followed by [1,3]-H shifts to give 2,2′-((1,4-dimethoxy-1,4-dioxobutane-2,3-diylidene)bis(azanylylidene))bis(quinoline-3-carboxylic acid) **3** in good yield. The structure of compound **3** was confirmed by spectroscopic characterization. Finally, compound **3** was proven to be significantly more stable than intermediate **A** as evidenced by DFT calculations.

Supplementary Materials: The following are available online. ^{1}H-NMR spectrum of compound **3** and HRMS Qualitative Compound Report.

Author Contributions: Conceptualization, J.F. and J.S.; Synthesis, J.F.; Molecule characterization, J.F. and P.W.; NMR data analysis, J.F.; DFT calculation and visualization, K.G. and J.S.; Data evaluation, J.F., K.G., P.W., and J.S.; Writing—original draft preparation, J.F.; Writing—review and editing, J.S.; Supervision and project administration, J.S. All authors read and approved the final manuscript.

Funding: This research was funded by the Funds for Statutory Activity of the Medical University of Gdansk (ST-020038/07).

Acknowledgments: The NMR spectrum was carried out at The Intercollegiate Nuclear Magnetic Resonance Laboratory, Gdańsk University of Technology, Poland. The elemental analysis and HRMS experiments were performed at The Physics-Chemistry Workshops, Faculty of Chemistry, University of Gdańsk, Poland. We thank Prof. Franciszek Sączewski, at the Medical University of Gdańsk, Poland, for his consultation on reaction mechanism.

References

1. Ergenc, N.; Capan, G.; Otuk, G. New 4-arylhydrazono-5(4*H*)-isoxazolone derivatives as possible antibacterial and anticonvulsant agents. *Pharmazie* **1993**, *48*, 780–782.

2. Mazimba, O.; Wale, K.; Loeto, D.; Kwape, T. Antioxidant and antimicrobial studies on fused-ring pyrazolones and isoxazolones. *Bioorg. Med. Chem.* **2014**, *22*, 6564–6569. [CrossRef]

3. Chande, M.S.; Verma, R.S.; Barve, P.A.; Khanwelkar, R.R.; Vaidya, R.B.; Ajaikumar, K.B. Facile synthesis of active antitubercular, cytotoxic and antibacterial agents: A Michael addition approach. *Eur. J. Med. Chem.* **2005**, *40*, 1143–1148. [CrossRef]

4. Yu, M.; Wang, J.; Tang, K.; Shi, X.; Wang, S.; Zhu, W.-M.; Zhang, X.-H. Purification and characterization of antibacterial compounds of *Pseudoalteromonas flavipulchra* JG1. *Microbiology* **2012**, *158*, 835–842. [CrossRef]

5. Wierenga, W.; Evans, B.R.; Zurenko, G.E. Benzisoxazolones: Antimicrobial and antileukemic activity. *J. Med. Chem.* **1984**, *27*, 1212–1215. [CrossRef]

6. Saidachary, G.; Veera Prasad, K.; Divya, D.; Singh, A.; Ramesh, U.; Sridhar, B.; China Raju, B. Convenient one-pot synthesis, anti-mycobacterial and anticancer activities of novel benzoxepinoisoxazolones and pyrazolones. *Eur. J. Med. Chem.* **2014**, *76*, 460–469. [CrossRef]

7. Padmavathi, V.; Venkata Subbaiah, D.R.C.; Mahesh, K.; Radha Lakshmi, T. Synthesis and Bioassay of Amino-pyrazolone, Amino-isoxazolone and Amino-pyrimidinone Derivatives. *Chem. Pharm. Bull.* **2007**, *55*, 1704–1709. [CrossRef]

8. Tanaka, K.; Matsuo, K.; Nakanishi, A.; Jo, M.; Shiota, H.; Yamaguchi, M.; Yoshino, S.; Kawaguchi, K. Syntheses and Antimicrobial Activities of Five-Membered Heterocycles Having a Phenylazo Substituent. *Chem. Pharm. Bull.* **1984**, *32*, 3291–3298. [CrossRef]

9. Braunholtz, J.T.; Freeman, P.F.H. Fungicidal Isoxazolones. GB1074803, 5 July 1967.

10. Das, N.P.; Mishra, P.K.; Sahu, S. Fungicidal activity of some substituted 5-isoxazolones. *Acta Cienc. Indica Chem.* **2011**, *37*, 239–243.

11. Tabarki, M.A.; Besbes, R. Regioselective ring opening of β-phenylglycidate and aziridine-2-carboxylates with *N*-alkylhydroxylamines: Synthesis of isoxazolidinones. *Tetrahedron* **2014**, *70*, 1060–1064. [CrossRef]

12. Heintz-Buschart, A.; Eickhoff, H.; Hohn, E.; Bilitewski, U. Identification of inhibitors of yeast-to-hyphae transition in *Candida albicans* by a reporter screening assay. *J. Biotechnol.* **2013**, *164*, 137–142. [CrossRef]

13. Miyake, T.; Yagasaki, Y.; Kagabu, S. Potential new fungicides: *N*-acyl-5-methyl-3(2*H*)-isoxazolone derivatives. *J. Pest. Sci.* **2012**, *37*, 89–94. [CrossRef]

14. Kömürcü, S.G.; Rollas, S.; Yilmaz, N.; Cevikbas, A. Synthesis of 3-methyl-4-[(2,4-dihydro-4-substituted-3*H*-1,2,4-triazole-3-thione-5-yl)phenylhydrazono]-5-isoxazolone and evaluation of their antimicrobial activities. *Drug Metabol. Drug Interact.* **1995**, *12*, 161–169. [CrossRef]

15. Summers, L.A. Agricultural fungicides. V. Preparation of 3-Methyl-4-(3′-pyridylhydrazono)-isoxazol-5-one and related compounds. *Aust. J. Chem.* **1973**, *26*, 2723–2725. [CrossRef]

16. Cox, M.; Jahangri, S.; Perkins, M.V.; Prager, R.H. Some Synthetic Approaches to Glutamate AMPA Receptor Agonists Based on Isoxazolones. *Aust. J. Chem.* **2004**, *57*, 685–688. [CrossRef]

17. Panathur, N.; Gokhale, N.; Dalimba, U.; Koushik, P.V.; Yogeeswari, P.; Sriram, D. New indole–isoxazolone derivatives: Synthesis, characterisation and in vitro SIRT1 inhibition studies. *Bioorg. Med. Chem. Lett.* **2015**, *25*, 2768–2772. [CrossRef]

18. Rollas, S.; Kokyan, Ş.; Koçyiğit-Kaymakçioğlu, B.; Özbaş-Turan, S.; Akbuğa, J. Synthesis and evaluation of cytotoxic activities of some substituted isoxazolone derivatives. *Marmara Pharm. J.* **2011**, *15*, 94–99. [CrossRef]

19. Vergelli, C.; Schepetkin, I.A.; Crocetti, L.; Iacovone, A.; Giovannoni, M.P.; Guerrini, G.; Khlebnikov, A.I.; Ciattini, S.; Ciciani, G.; Quinn, M.T. Isoxazol-5(2*H*)-one: A new scaffold for potent human neutrophil elastase (HNE) inhibitors. *J. Enzyme Inhib. Med.* **2017**, *32*, 821–831. [CrossRef]

20. Hou, X.Q.; Gao, Y.W.; Yang, S.T.; Wang, C.Y.; Ma, Z.Y.; Xia, X.Z. Role of macrophage migration inhibitory factor in influenza H5N1 virus pneumonia. *Acta Virol.* **2009**, *53*, 225–231. [CrossRef]

21. Laughlin, S.K.; Clark, M.P.; Djung, J.F.; Golebiowski, A.; Brugel, T.A.; Sabat, M.; Bookland, R.G.; Laufersweiler, M.J.; VanRens, J.C.; Townes, J.A.; et al. The development of new isoxazolone based inhibitors of tumor necrosis factor-alpha (TNF-α) production. *Bioorg. Med. Chem. Lett.* **2005**, *15*, 2399–2403. [CrossRef]

22. Deng, B.-L.; Hartman, T.L.; Buckheit, R.W., Jr.; Pannecouque, C.; De Clercq, E.; Cushman, M. Replacement of the Metabolically Labile Methyl Esters in the Alkenyldiarylmethane Series of Non-Nucleoside Reverse Transcriptase Inhibitors with Isoxazolone, Isoxazole, Oxazolone, or Cyano Substituents. *J. Med. Chem.* **2006**, *49*, 5316–5323. [CrossRef]

23. Ishioka, T.; Tanatani, A.; Nagasawa, K.; Hashimoto, Y. Anti-Androgens with full antagonistic activity toward human prostate tumor LNCaP cells with mutated androgen receptor. *Bioorg. Med. Chem. Lett.* **2003**, *13*, 2655–2658. [CrossRef]

24. Ishioka, T.; Kubo, A.; Koiso, Y.; Nagasawa, K.; Itai, A.; Hashimoto, Y. Novel Non-Steroidal/Non-Anilide Type Androgen Antagonists with an Isoxazolone Moiety. *Bioorg. Med. Chem.* **2002**, *10*, 1555–1566. [CrossRef]

25. Laufer, S.A.; Margutti, S. Isoxazolone Based Inhibitors of p38 MAP Kinases. *J. Med. Chem.* **2008**, *51*, 2580–2584. [CrossRef]

26. Demers, J.P.; Hageman, W.E.; Johnson, S.G.; Klaubert, D.H.; Look, R.A.; Moore, J.B. Selective inhibitors of protein kinase C in a model of graft-*vs*-host disease. *Bioorg. Med. Chem. Lett* **1994**, *4*, 2451–2456. [CrossRef]

27. Kafle, B.; Aher, N.G.; Khadka, D.; Park, H.; Cho, H. Isoxazol-5(4*H*)one Derivatives as PTP1B Inhibitors Showing an Anti-Obesity Effect. *Chem. Asian J.* **2011**, *6*, 2073–2079. [CrossRef]

28. Kafle, B.; Cho, H. Isoxazolone Derivatives as Potent Inhibitors of PTP1B. *Bull. Korean Chem. Soc.* **2012**, *33*, 275–277. [CrossRef]

29. Frolund, B.; Kristiansen, U.; Brehm, L.; Hansen, A.B.; Krogsgaard-Larsen, P.; Falch Partial, E. Partial GABA$_A$ Receptor Agonists. Synthesis and in Vitro Pharmacology of a Series of Nonannulated Analogs of 4,5,6,7-Tetrahydroisoxazolo[4,5-*c*]pyridin-3-ol. *J. Med. Chem.* **1995**, *38*, 3287–3296. [CrossRef]

30. Tamura, N.; Matsushita, Y.; Kishimoto, S.; Itoh, K. Synthesis and Biological Activity of (S)-2-Amino-3-(2, 5-dihydro-5-oxo-4-isoxazolyl)propanoic Acid (TAN-950 A) Derivatives. *Chem. Pharm. Bull.* **1991**, *39*, 1199–1212. [CrossRef]

31. Tamura, N.; Itoh, K.; Iwama, T. Synthesis and Glutamate-Agonistic Acitivity of (S)-2-Amino-3-(2, 5-dihydro-5-oxo-3-isoxazolyl)-propanoic Acid Derivatives. *Chem. Pharm. Bull.* **1992**, *40*, 381–386. [CrossRef]

32. Toshi, I.; Yasuo, N.; Norikazu, T.; Setsuo, H.; Akinobu, N. A novel glutamate agonist, TAN-950 A, isolated from streptomycetes. *Eur. J. Pharmacol.* **1991**, *197*, 187–192. [CrossRef]

33. Shinabarger, D. Mechanism of action of the oxazolidinone antibacterial agents. *Exp. Opin. Investig. Drugs* **1999**, *8*, 1195–1202. [CrossRef] [PubMed]

34. Sączewski, J.; Jalińska, A.; Kędzia, A. New derivatives of 4,6-dimethylisoxazolo[3,4-*b*] pyridin-3(1*H*)-one: Synthesis, tautomerism, electronic structure and antibacterial activity. *Heterocycl. Commun* **2014**, *20*, 215–223. [CrossRef]

35. Sączewski, J.; Fedorowicz, J.; Kędzia, A.; Ziółkowska-Klinkosz, M.; Jalińska, A. Synthesis and Antifungal Activity of Some 4,6-Dimethylisoxazolo[3,4-*b*]pyridin-3(1*H*)-one Derivatives. *Med. Chem.* **2016**, *12*, 640–646. [CrossRef]

36. Fedorowicz, J.; Sączewski, J.; Konopacka, A.; Waleron, K.; Lejnowski, D.; Ciura, K.; Tomašič, T.; Skok, Ž.; Savijoki, K.; Morawska, M.; et al. Synthesis and biological evaluation of hybrid quinolone-based quaternary ammonium antibacterial agents. *Eur. J. Med. Chem.* **2019**, *179*, 576–590. [CrossRef]

37. Ciura, K.; Fedorowicz, J.; Andrić, F.; Greber, K.E.; Gurgielewicz, A.; Sawicki, W.; Sączewski, J. Lipophilicity Determination of Quaternary (Fluoro)Quinolones by Chromatographic and Theoretical Approaches. *Int. J. Mol. Sci.* **2019**, *20*, 5288. [CrossRef]

38. Sączewski, J.; Fedorowicz, J.; Korcz, M.; Sączewski, F.; Wicher, B.; Gdaniec, M.; Konopacka, A. Experimental and theoretical studies on the tautomerism and reactivity of isoxazolo[3,4-*b*]quinolin-3(1*H*)-ones. *Tetrahedron* **2015**, *71*, 8975–8984. [CrossRef]

39. Sączewski, J.; Fedorowicz, J.; Gdaniec, M.; Wiśniewska, P.; Sieniawska, E.; Drażba, Z.; Rzewnicka, J.; Balewski, Ł. The Elusive Paal–Knorr Intermediates in the Trofimov Synthesis of Pyrroles: Experimental and Theoretical Studies. *J. Org. Chem.* **2017**, *82*, 9737–9743. [CrossRef]

40. Sączewski, J.; Hinc, K.; Obuchowski, M.; Gdaniec, M. The Tandem Mannich–Electrophilic Amination Reaction: A Versatile Platform for Fluorescent Probing and Labeling. *Chem. Eur. J.* **2013**, *19*, 11531–11535. [CrossRef]

41. Fedorowicz, J.; Sączewski, J.; Drażba, Z.; Wiśniewska, P.; Gdaniec, M.; Wicher, B.; Suwiński, G.; Jalińska, A. Synthesis and fluorescence of dihydro-[1,2,4]triazolo[4,3-*a*]pyridin-2-iumcarboxylates: An experimental and TD-DFT comparative study. *Dyes Pigments* **2019**, *161*, 347–359. [CrossRef]

42. Frisch, M.J.; Trucks, G.W.; Schlegel, H.B.; Scuseria, G.E.; Robb, M.A.; Cheeseman, J.R.; Scalmani, G.; Barone, V.; Mennucci, B.; Petersson, G.A.; et al. *Gaussian 16, Revision A.03*; Gaussian, Inc.: Wallingford, CT, USA, 2016.

43. Lee, C.; Yang, W.; Parr, R.G. Development of the Colle-Salvetti correlation-energy formula into a functional of the electron density. *Phys. Rev. B* **1988**, *37*, 785–789. [CrossRef] [PubMed]

44. Chai, J.-D.; Head-Gordon, M. Systematic optimization of long-range corrected hybrid density functionals. *J. Chem. Phys.* **2008**, *128*, 084106. [CrossRef] [PubMed]

45. Austin, A.; Petersson, G.A.; Frisch, M.J.; Dobek, F.J.; Scalmani, G.; Throssell, K. A density functional with spherical atom dispersion terms. *J. Chem. Theor. Comput.* **2012**, *8*, 4989–5007. [CrossRef] [PubMed]

46. Tomasi, J.; Mennucci, B.; Cammi, R. Quantum mechanical continuum solvation models. *Chem. Rev.* **2005**, *105*, 2999–3094. [CrossRef]

Short Note

(1*R*,4*S*,5*S*)-5-((3-Hydroxypropyl)amino)-4-((1-methyl-1*H*-tetrazol-5-yl)thio)cyclopent-2-en-1-ol

Milene A. G. Fortunato, Filipa Siopa * and Carlos A. M. Afonso *

Research Institute for Medicines (iMed.ULisboa), Faculty of Pharmacy, Universidade de Lisboa,
Av. Prof. Gama Pinto, 1649-003 Lisboa, Portugal; milene.fortunato@campus.ul.pt
* Correspondence: filipasiopa@ff.ulisboa.pt (F.S.); carlosafonso@ff.ulisboa.pt (C.A.M.A.);
  Tel.: +351-21-7946400 (ext. 321) (F.S. & C.A.M.A.)

Abstract: Using environmentally friendly conditions, the nucleophilic ring-opening reaction of 6-azabicyclo[3.1.0]hex-3-en-2-ol with 1-methyl-1*H*-tetrazole-5-thiol provided a novel thiol-incorporated aminocyclopentitol, (1*R*,4*S*,5*S*)-5-((3-hydroxypropyl)amino)-4-((1-methyl-1*H*-tetrazol-5-yl)thio) cyclopent-2-en-1-ol, in excellent yield (95%). The newly synthesized compound was analyzed and characterized via ^{1}H, ^{13}C-NMR, HSQC, and mass spectral data.

Keywords: aminocyclopentitol; bicyclic aziridine; water chemistry; nucleophilic substitution

Citation: Fortunato, M.A.G.; Siopa, F.; Afonso, C.A.M. (1*R*,4*S*,5*S*)-5-((3-Hydroxypropyl)amino)-4-((1-methyl-1*H*-tetrazol-5-yl)thio)cyclopent-2-en-1-ol. *Molbank* **2021**, *2021*, M1199. https://doi.org/10.3390/M1199

Academic Editors: Dimitrios Matiadis and Eleftherios Halevas

Received: 14 March 2021
Accepted: 27 March 2021
Published: 1 April 2021

Publisher's Note: MDPI stays neutral with regard to jurisdictional claims in published maps and institutional affiliations.

1. Introduction

Aziridines are recurrent motifs in anticancer compounds (Figure 1A) and are useful building blocks in organic synthesis, largely due to their ring strain [1].

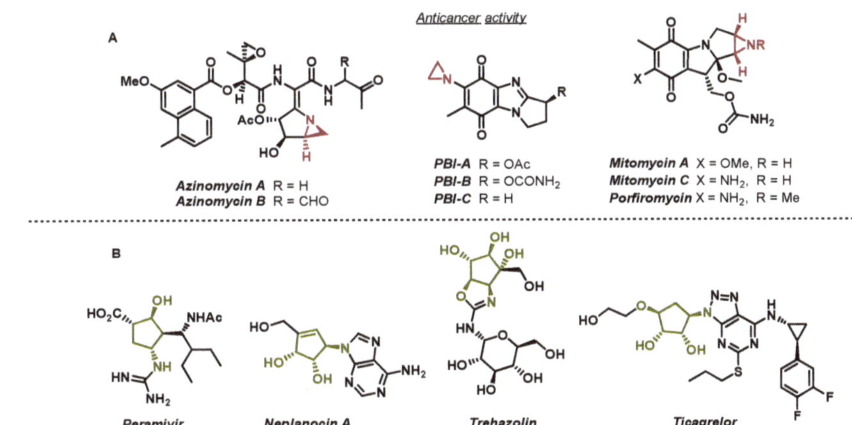

Figure 1. Examples of chemical structures of bioactive compounds containing aziridines (**A**) and important biological compounds containing the aminocyclopentitol scaffold (**B**) [2–4].

In 1972, Kaplan and co-workers implemented an innovative methodology to prepare 6-azabicyclo[3.1.0]hex-3-en-2-ol (bicyclic vinyl aziridines) via the photochemical conversion of pyridinium salts (Scheme 1) [5]. Bicyclic vinyl aziridines are useful intermediates to access aminocyclopentitols, a family of natural compounds known for being glycosidase inhibitors (Figure 1B) [6,7]. The synthetic methodology to prepare aminocyclopentitols involves the synthesis of bicyclic vinyl aziridines, followed by a ring-opening reaction to originate an aminocyclopentene, which after further functionalization originate the desired aminocyclopentitol. This methodology was developed by Mariano [8] and applied to the synthesis of several aminocyclopentitols, such as (+)-mannostatin A [9], (+)-castanospermine [10], and (-)-swainsonine [11] (Scheme 1).

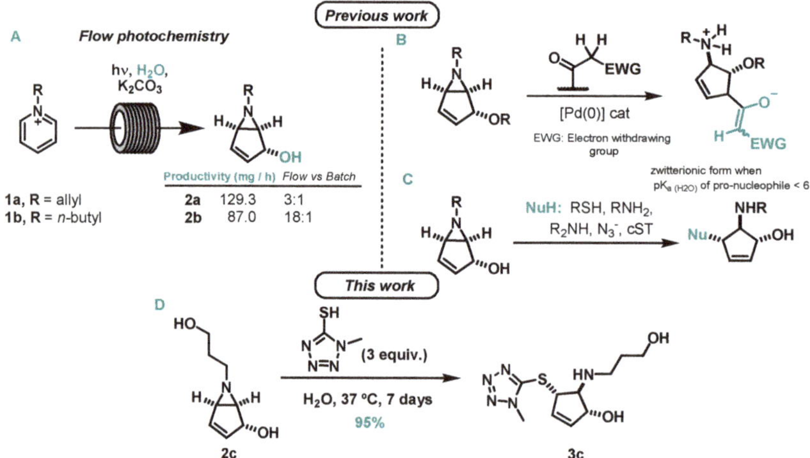

Scheme 1. Synthetic methodology to prepare aminocyclopentitols by taking advantage of bicyclic vinyl aziridines prepared via photoreactions of pyridinium salts [9–11].

Mannostatin A is a natural product, first isolated from a soil microorganism *Streptoverticillus*, and is among the most potent inhibitors of class II α-mannosidase. The chemical structure of Mannostatin A contains a thiol functionality, responsible for the high affinity to the enzyme's binding site [12]. Inhibitors of glycosidases are leading the drug discovery across cancer, and viral and bacterial infections. Mannostatin A and its analogs [13] have been used to study the inhibition of glycosidases to orientate the development of drug candidates [6,7].

Our group studied the photochemical reactions of pyridinium salts to bicyclic vinyl aziridines under continuous-flow [14,15]. The implementation of flow enabled the synthesis of bicyclic vinyl aziridines with larger productivity when compared to reported batch methods [16] and also the achievement of a gram scale production [15] (Scheme 2A).

Scheme 2. (**A**) Flow photocyclization of pyridinium salts [14,15], improvements over batch methodology [16,17]; previous work on (**B**) palladium-catalyzed allylic substitutions on bicyclic vinyl aziridines using carbon-based nucleophiles [18] and (**C**) nucleophilic ring-opening reactions of bicyclic vinyl aziridines with thiol and nitrogen nucleophiles [17]; (**D**) this work: nucleophilic ring-opening reaction of bicyclic vinyl aziridine ring (**2c**) by 1-methyl-1H-tetrazole-5-thiol to produce **3c**.

We also studied several bicyclic vinyl aziridines transformations. In collaboration with G. Poli, we reported a palladium-catalyzed allylic substitutions using C-nucleophiles [18] (Scheme 2B). Additionally, we accomplished several ring-opening reactions using sulfur and nitrogen-based nucleophiles in an aqueous medium, including a bioconjugation with the peptide hormone salmon calcitonin (sCT) [17] (Scheme 2C). Within the reactions performed, the best yields were achieved for thiol-nucleophiles. In line with this work, we herein present the ring-opening reaction of 6-(3-hydroxypropyl)-6-azabicyclo[3.1.0]hex-3-en-2-ol by a thiol-based nucleophile, 1-methyl 1*H* tetrazole-5-thiol (Scheme 2D).

2. Results and Discussion

The starting material, 1-(3-hydroxypropyl)pyridin-1-ium chloride (**1c**), was prepared from pyridine and 3-chloropronanol following our reported method. 6-(3-hydroxypropyl)-6-azabicyclo[3.1.0]hex-3-en-2-ol (**2c**) was obtained by photohydration of the pyridinium salt **1c** (Scheme 3), also taking advantage of our previous reported synthetic methodology [18]. In this work, the bicyclic vinyl aziridine **2c** was subjected to thiol-nucleophilic attack by 1-methyl-1*H*-tetrazole-5-thiol, applying our reported optimized conditions [17]. An excess of the nucleophile (3 equivalents) was used under mild reaction conditions (37 °C, in water), and we obtained the product **3c** as a brown oil, in excellent yield (95%) after purification by silica gel chromatography (Scheme 3). The ring-opening reaction occurs via the S_N2 pathway in a regio- and stereospecific manner, and the nucleophile attacks in the less sterically hindered carbon of the aziridine moiety, as previously reported by Mariano [19] and Burger [20], and more recently also by us [17].

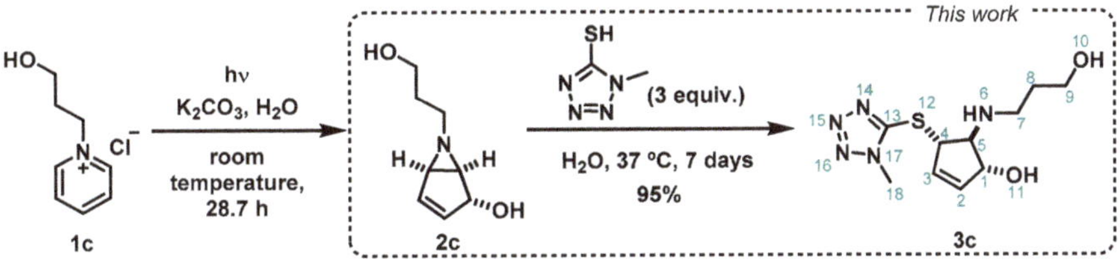

Scheme 3. Synthetic pathway from photocyclization of 1-(3-hydroxypropyl)pyridin-1-ium chloride (1c) [18], followed by a nucleophilic attack to the bicyclic vinyl aziridine ring (**2c**) by 1-methyl-1H-tetrazole-5-thiol to produce **3c**.

The product **3c** was characterized by ^{1}H-NMR, ^{13}C-NMR, HSQC, and HRMS. By analyzing the ^{1}H-NMR spectrum (Figure S1), we can observe characteristic peaks from product **3c**: a singlet at 3.96 ppm, corresponding to the methyl group linked to the tetrazole ring (H-18), and multiplets corresponding to the geminal protons of the thioether at 4.30–4.29 ppm (H-4), the alcohol at 4.59–4.58 ppm (H-1), and the amine at 3.29–3.26 ppm (H-5). The signals for the hydroxypropyl chain can be observed as a quartet (J = 6.9 Hz, 2.77 ppm), a multiplet (1.76–1.67 ppm), and a triplet (J = 6.4 Hz, 3.60 ppm), corresponding to the protons vicinal to the amine (H-7), to the middle-chain methylene (H-8), and to the protons geminal to the hydroxyl group (H-9), respectively. Additionally, the ^{13}C-NMR (Figure S2) shows the characteristic peaks from the tetrazole ring: a quaternary carbon at 153.32 ppm (C-13), which does not correlate with a proton signal in the HSQC (Figure S3), and the carbon from the methyl group at 33.87 ppm (C-18).

The product **3c** can be further functionalized, since it has a primary and a secondary hydroxyl group. Moreover, **3c** has a tetrazole ring which could lead to potential biological activity, since the tetrazole moiety can be found in different approved [21] and candidate drugs [22,23]. Accomplishing the synthesis of **3c** contributed to expanding our previous aminocyclopentitols library [17].

3. Materials and Methods

All chemicals, reagents, and solvents were of analytical grade, purchased from commercial sources, namely, Merck (Algés, Portugal) and Alfa Aesar (Kandel, Germany) and were used without further purification. NMR spectra were obtained on a Bruker Fourier 300 spectrometer (Bruker BioSpin AG, Fallanden, Switzerland) using TopSpin® software (Bruker BioSpin GmbH, Rheinstetten, Germany). NMR experiments were performed in D_2O at room temperature. Chemical shifts are given in parts per million (ppm); the terms m, s, d, t, and q represent multiplet, singlet, doublet, triplet, and quartet, respectively; and the coupling constants (J) are given in Hertz (Hz). High-resolution mass spectroscopy (HRMS) was performed in a LTQ Orbitrap XL mass spectrometer, Thermo Fischer Scientific, Bremen, Germany.

1-(3-Hydroxypropyl)pyridin-1-ium chloride (**1c**) and (*1R,2R,5R*)-6-(3-hydroxypropyl)-6-azabicyclo[3.1.0]hex-3-en-2-ol (**2c**) were prepared as previously described by us [18].

(1R,4S,5S)-5-((3-hydroxypropyl)amino)-4-((1-methyl-1H-tetrazol-5-yl)thio)cyclopent-2-en-1-ol **3c**: To a solution of (*1R,2R,5R*)-6-(3-hydroxypropyl)-6-azabicyclo[3.1.0]hex-3-en-2-ol **2c** (29.6 mg; 0.19 mmol) in distilled water (1 mL), 1-methyl-1H-tetrazole-5-thiol (60.1 mg; 0.57 mmol, 3 equiv.) was added. The reaction mixture was stirred at 37 °C and followed by TLC (eluent: dichloromethane/methanol, 9:1) until the complete disappearance of the starting material, observed after 7 days. The crude reaction was concentrated under reduced pressure and purified by silica gel chromatography eluting with dichloromethane, methanol, and triethylamine (9:1:0.1) to afford the ring-opening product **3c** as a brown oil in 95% yield (49.23 mg).

^{1}H-NMR (300 MHz, D_2O) δ 5.95–5.90 (m, 2H, H-2 and H-3), 4.59–4.58 (m, 1H, H-1), 4.30–4.29 (m, 1H, H-4), 3.96 (s, 3H, H-18), 3.60 (t, J = 6.4 Hz, 2H, H-9), 3.29–3.26 (m, 1H, H-5), 2.77 (q, J = 6.9 Hz, 2H, H-7), 1.76-1.67 (m, 2H, H-8).

^{13}C-NMR (100 MHz, D_2O) δ 153.32 (C-13), 135.02 (C-3), 131.91 (C-2), 80.01 (C-1), 72.44 (C-5), 59.60 (C-9), 55.55 (C-4), 44.05 (C-7), 33.87 (C-18), 30.67 (C-8).

HRMS m/z calc. for $C_{10}H_{17}N_5O_2S$ [M + H]$^+$ 272.11757, obtained 272.11740.

4. Conclusions

We obtained (*1R,4S,5S*)-5-((3-hydroxypropyl)amino)-4-((1-methyl-1H-tetrazol-5-yl)thio) cyclopent-2-en-1-ol (**3c**) through bicyclic vinyl aziridine ring-opening reaction, with 1-methyl-1H-tetrazole-5-thiol. The reaction was executed under mild (37 °C) and sustainable (water as reaction medium) conditions. The compound **3c** was characterized using 1H NMR, 13C NMR, HSQC, and HRMS.

Supplementary Materials: The following are available online, Figure S1: ^{1}H NMR spectrum; Figure S2: ^{13}C-NMR spectrum; Figure S3: HSQC spectrum; Figure S4: HRMS.

Author Contributions: Conceptualization and supervision, F.S. and C.A.M.A.; investigation, M.A.G.F.; data curation and formal analysis, M.A.G.F. and F.S.; writing, M.A.G.F., F.S. and C.A.M.A. All authors have read and agreed to the published version of the manuscript.

Funding: This research was funded by Fundação para a Ciência e a Tecnologia (FCT) (reference number: PTDC/QUI-QOR/32008/2017). The project leading to this application received funding from the European Union's Horizon 2020 research and innovation program under grant agreement No. 951996.

Institutional Review Board Statement: Not applicable.

Informed Consent Statement: Not applicable.

Data Availability Statement: Not applicable.

References

1. Chai, Z. Catalytic Asymmetric Transformations of Racemic Aziridines. *Synthesis* **2020**, *52*, 1738–1750. [CrossRef]
2. Sweeney, J.B. Aziridines: Epoxides' ugly cousins? *Chem. Soc. Rev.* **2002**, *31*, 247–258. [CrossRef]
3. Košak, U.; Hrast, M.; Knez, D.; Maraš, N.; Črnugelj, M.; Gobec, S. Convenient syntheses of orthogonally protected aminocyclopentitols from aldopentoses. *Tetrahedron Lett.* **2015**, *56*, 529–531. [CrossRef]
4. Marin, L.; Force, G.; Gandon, V.; Schulz, E.; Lebœuf, D. Aza-Piancatelli Cyclization as a Platform for the Preparation of Scaffolds of Natural Compounds: Application to the Total Synthesis of Bruceolline D. *Eur. J. Org. Chem.* **2020**, *2020*, 5323–5328. [CrossRef]
5. Kaplan, L.; Pavlik, J.W.; Wilzbach, K.E. Photohydrataion of Pyridinium Ions. *J. Am. Chem. Soc.* **1972**, *94*, 3283–3284. [CrossRef]
6. Delgado, A. Recent Advances in the Chemistry of Aminocyclitols. *Eur. J. Org. Chem.* **2008**, *2008*, 3893–3906. [CrossRef]
7. Diaz, L.; Delgado, A. Medicinal Chemistry of Aminocyclitols. *Curr. Med. Chem.* **2010**, *17*, 2393–2418. [CrossRef]
8. Yoon, U.C.; Quillen, S.L.; Mariano, P.S.; Swanson, R.; Stavinoha, J.L.; Bay, E. Electron transfer initiated photocyclizations of N-allylpridinium and quinolinium salts. *Tetrahedron Lett.* **1982**, *23*, 919–922. [CrossRef]
9. Ling, R.; Mariano, P.S. A Demonstration of the Synthetic Potential of Pyridinium Salt Photochemistry by Its Application to a Stereocontrolled Synthesis of (+)-Mannostatin A 1. *J. Org. Chem.* **1998**, *63*, 6072–6076. [CrossRef] [PubMed]
10. Zhao, Z.; Song, L.; Mariano, P.S. A concise sequential photochemical-metathesis approach for the synthesis of (+)-castanospermine and possible uniflorine-A stereoisomers. *Tetrahedron* **2005**, *61*, 8888–8894. [CrossRef]
11. Song, L.; Duesler, E.N.; Mariano, P.S. Stereoselective synthesis of polyhydroxylated indolizidines based on pyridinium salt photochemistry and ring rearrangement metathesis. *J. Org. Chem.* **2004**, *69*, 7284–7293. [CrossRef] [PubMed]
12. Kawatkar, S.P.; Kuntz, D.A.; Woods, R.J.; Rose, D.R.; Boons, G.J. Structural basis of the inhibition of Golgi α-mannosidase II by mannostatin A and the role of the thiomethyl moiety in ligand-protein interactions. *J. Am. Chem. Soc.* **2006**, *128*, 8310–8319. [CrossRef] [PubMed]
13. Cho, S.J.; Ling, R.; Kim, A.; Mariano, P.S. A Versatile Approach to the Synthesis of (+) -Mannostatin A Analogues. *J. Org. Chem.* **2000**, *65*, 1574–1577. [CrossRef] [PubMed]
14. Siopa, F.; António, J.P.M.; Afonso, C.A.M. Flow-Assisted Synthesis of Bicyclic Aziridines via Photochemical Transformation of Pyridinium Salts. *Org. Process Res. Dev.* **2018**, *22*, 551–556. [CrossRef]
15. Fortunato, M.A.G.; Ly, C.-P.; Siopa, F.; Afonso, C.A.M. Process Intesification for the Synthesis of 6-Allyl-6-azabicyclo[3.1.0]hex-3-en-2-ol from 1-Allylpyridinium Salt Using a Continuous UV-Light Photoflow Approach. *Methods Protoc.* **2019**, *2*, 67. [CrossRef]
16. Colombo, C.; Pinto, B.M.; Bernardi, A.; Bennet, A.J. Synthesis and evaluation of influenza A viral neuraminidase candidate inhibitors based on a bicyclo[3.1.0]hexane scaffold. *Org. Biomol. Chem.* **2016**, *14*, 6539–6553. [CrossRef]
17. Vale, J.R.; Siopa, F.; Branco, P.S.; Afonso, C.A.M. Ring Opening of 6-Azabicyclo-[3.1.0]hex-3-en-2-ols in Water under Mild Conditions. *Eur. J. Org. Chem.* **2016**, *2016*, 2048–2053. [CrossRef]
18. Oliveira, J.A.C.; Kiala, G.; Siopa, F.; Bernard, A.; Gontard, G.; Oble, J.; Afonso, C.A.M.; Poli, G. Palladium-catalyzed allylic substitution between C-based nucleophiles and 6-azabicyclo[3.1.0]-hex-3-en-2-oxy derivatives: A new selectivity paradigm. *Tetrahedron* **2020**, *76*, 131182. [CrossRef]
19. Ling, R.; Yoshida, M.; Mariano, P.S. Exploratory investigations probing a preparatively versatile, pyridinium salt photoelectrocyclization-solvolytic aziridine ring opening sequence. *J. Org. Chem.* **1996**, *61*, 4439–4449. [CrossRef]
20. Acar, E.A.; Glarner, F.; Burger, U. Aminocyclopentitols from N-Alkylpyridinium Salts: A photochemical approach. *Helv. Chim. Acta* **1998**, *81*, 1095–1104. [CrossRef]
21. Gleiter, C.H.; Jägle, C.; Gresser, U.; Mörike, K. Candesartan. *Cardiovasc. Drug Rev.* **2004**, *22*, 263–284. [CrossRef] [PubMed]
22. Warrilow, A.G.S.; Hull, C.M.; Parker, J.E.; Garvey, E.P.; Hoekstra, W.J.; Moore, W.R.; Schotzinger, R.J.; Kelly, D.E.; Kellya, S.L. The clinical candidate VT-1161 is a highly potent inhibitor of candida albicans CYP51 but fails to bind the human enzyme. *Antimicrob. Agents Chemother.* **2014**, *58*, 7121–7127. [CrossRef] [PubMed]
23. Lockhart, S.R.; Fothergill, A.W.; Iqbal, N.; Bolden, C.B.; Grossman, N.T.; Garvey, E.P.; Brand, S.R.; Hoekstra, W.J.; Schotzinger, R.J.; Ottinger, E.; et al. The Investigational Fungal Cyp51 Inhibitor VT-1129 Demonstrates Potent In Vitro Activity against Cryptococcus neoformans and Cryptococcus gattii. *Antimicrob. Agents Chemother.* **2016**, *60*, 2528–2531. [CrossRef] [PubMed]

Communication

(*E*)-(1-(4-Ethoxycarbonylphenyl)-5-(3,4-dimethoxyphenyl)-3-(3,4-dimethoxystyryl)-2-pyrazoline: Synthesis, Characterization, DNA-Interaction, and Evaluation of Activity Against Drug-Resistant Cell Lines

Dimitris Matiadis [1], Barbara Mavroidi [1], Angeliki Panagiotopoulou [1], Constantinos Methenitis [2], Maria Pelecanou [1] and Marina Sagnou [1,*

[1] National Center for Scientific Research "Demokritos", Institute of Biosciences & Applications, 153 10 Athens, Greece; matiadis@bio.demokritos.gr (D.M.); bmavroidi@bio.demokritos.gr (B.M.); apanagio@bio.demokritos.gr (A.P.); pelmar@bio.demokritos.gr (M.P.)

[2] Department of Chemistry, National and Kapodistrian University of Athens, 157 84 Athens, Greece; methenitis@chem.uoa.gr

* Correspondence: sagnou@bio.demokritos.gr; Tel.: +30-210-6503558

Received: 15 January 2020; Accepted: 27 January 2020; Published: 30 January 2020

Abstract: (*E*)-1-(4-Ethoxycarbonylphenyl)-5-(3,4-dimethoxyphenyl)-3-(3,4-dimethoxystyryl)-2-pyrazoline was synthesized via the cyclization reaction between the monocarbonyl curcuminoid (2*E*,6*E*)-2,6-bis(3,4-dimethoxybenzylidene)acetone and ethyl hydrazinobenzoate in high yield and purity (>95% by High-performance liquid chromatography (HPLC)). The compound has been fully characterized by ^{1}H, ^{13}C NMR, FTIR, UV-Vis and HRMS and its activity was evaluated in terms of its potential interaction with DNA as well as its cytotoxicity against resistant and non-resistant tumor cells. Both DNA thermal denaturation and DNA viscosity measurements revealed that a significant intercalation binding takes place upon treatment of the DNA with the synthesized pyrazoline, causing an increase in melting temperature by 3.53 ± 0.11 °C and considerable DNA lengthening and viscosity increase. However, neither re-sensitisation of Doxorubicin (DO X)-resistant breast cancer and multidrug resistance (MDR) reversal nor synergistic activity with DOX by potentially increasing the DOX cell killing ability was observed.

Keywords: pyrazolines; curcuminoids; nitrogen heterocycles; cytotoxic; DNA binding; MDR reversal

---

1. Introduction

Nitrogen-containing heterocycles are important core structures found in many natural products [1] and synthetic compounds exhibiting a broad range of biological activities. Recent analysis of a Food and Drug Administration-approved drug database revealed that 59% of unique small-molecule drugs contain at least one nitrogen heterocycle [2]. Pyrazolines in particular, constitute a class of five-membered heterocycles incorporating a nitrogen-nitrogen (N-N) bond [3], with remarkable pharmacological applications [4] as anticancer [5–9], anti-inflammatory [10–12], antimicrobial [13,14], and antimalarial drugs [15].

Representative examples of synthetic bioactive pyrazolines are depicted in Figure 1. Thiazolone containing pyrazoline **1** has been found to be selectively active against colon cancer cell lines, especially on HT-29 [6]. Benzimidazole pyrazoline **2** has been reported by Shaharyar et al. as the most active antitumor compound among a library of selected similar derivatives [7], whereas benzenesulfonamide pyrazoline **3** was shown by Rathish et al. in in vitro and in vivo studies to be a more potent anti-inflammatory agent than celecoxib [10]. Manna et al. investigated 1-acetyl-3,5-diaryl-pyrazolines **4a** and **4b** for their anticancer activity and binding affinity to P-glycoprotein [16]. On the other hand, 3,5-divinylpyrazole **5** (Figure 1) which is a curcumin derivative, has been reported by Kolotova et al. as a potent inhibitor of P-glycoprotein, which is associated with the induction of multidrug resistance in cancer chemotherapy [17,18]. Such molecules have consequently caused resensitisation of the resistant cells and reversal of the multidrug resistance phenomenon [18]. Taking into account the aforementioned promising findings, we synthesized compound **6** (Figure 2) which is the pyrazoline analogue of the curcuminoid pyrazole derivative **5**, and we studied its interaction with DNA as well as its cytotoxicity against chemo-resistant and non-resistant tumor cells.

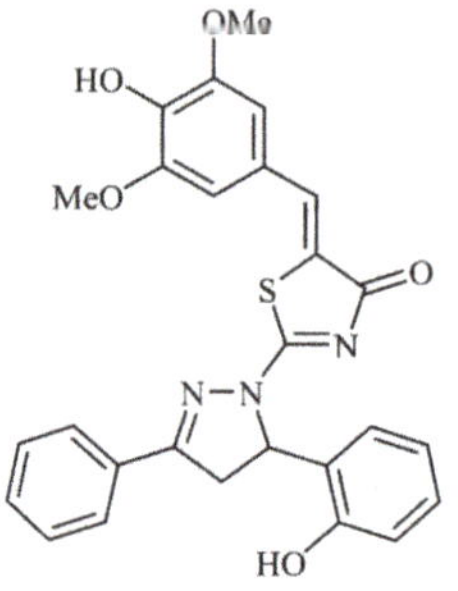

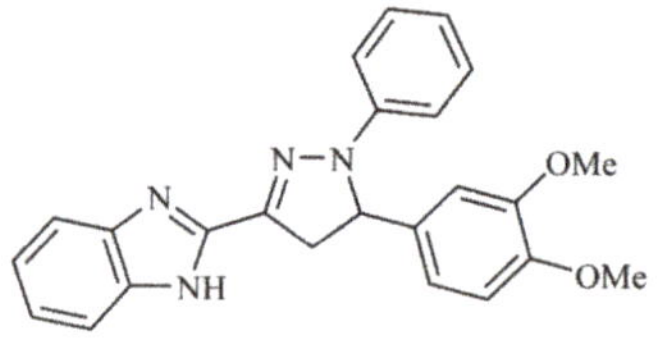

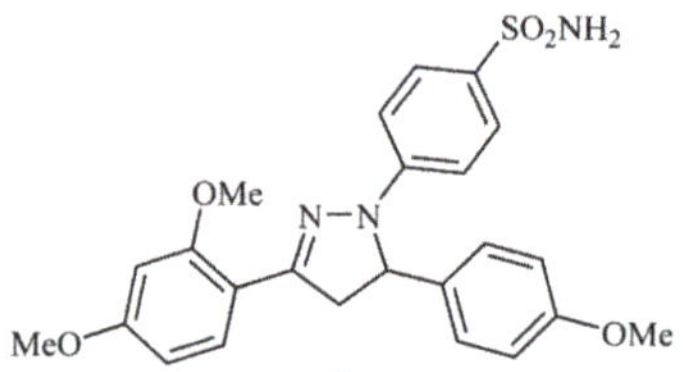

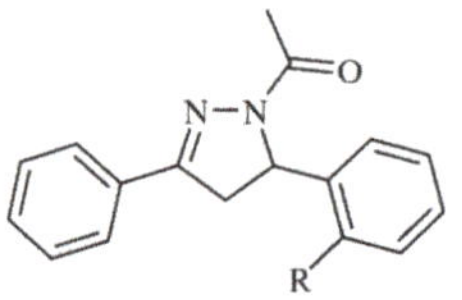

Figure 1. Structures of known biologically active pyrazolines **1–4**, pyrazole **5** and title pyrazoline **6**. The numbering of compound **6** skeleton is according to common pyrazoline nomenclature. The product **6** is racemic, however the relative configuration is shown for clarity and simplicity purposes to describe the AMX system and the assignments.

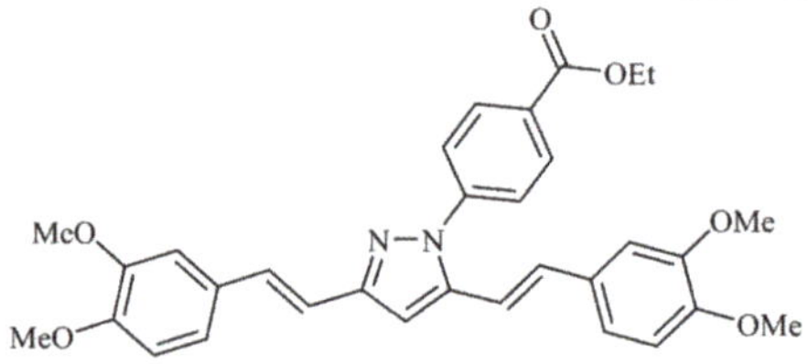

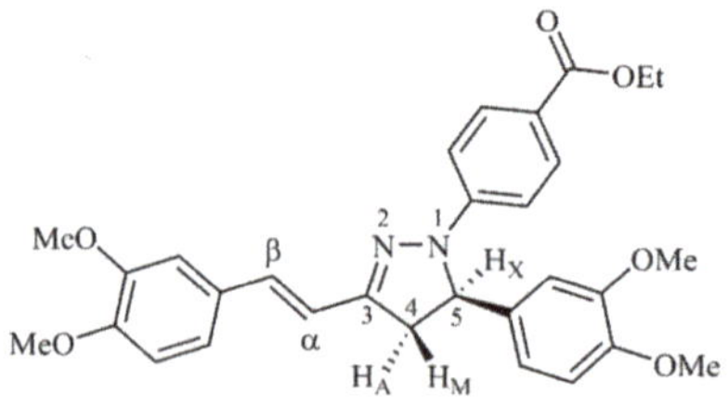

2. Results and Discussion

2.1. Synthesis and Characterization

The synthetic route towards the preparation of the title compound **6** is outlined in Schemes 1 and 2. In the first step, the intermediate symmetrical monocarbonyl curcuminoid **7** was synthesized from acetone and veratraldehyde via base-catalyzed Claisen–Schmidt condensation reaction and used after recrystallization from ethanol (Scheme 1).

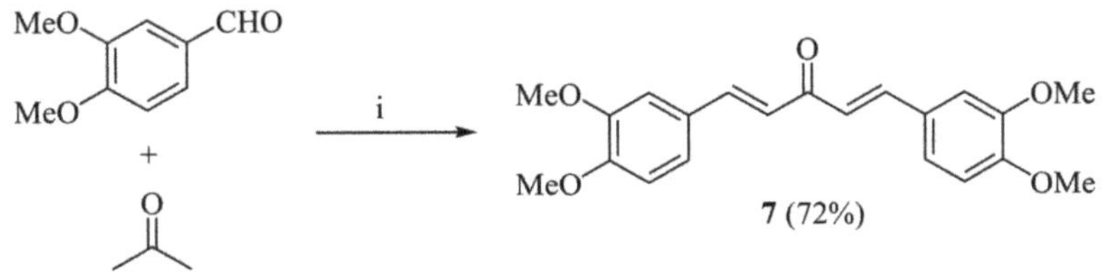

Scheme 1. Synthesis of monocarbonyl curcuminoid intermediate **7**. Reagents and conditions: (i) NaOH, EtOH, 2 h, r.t.

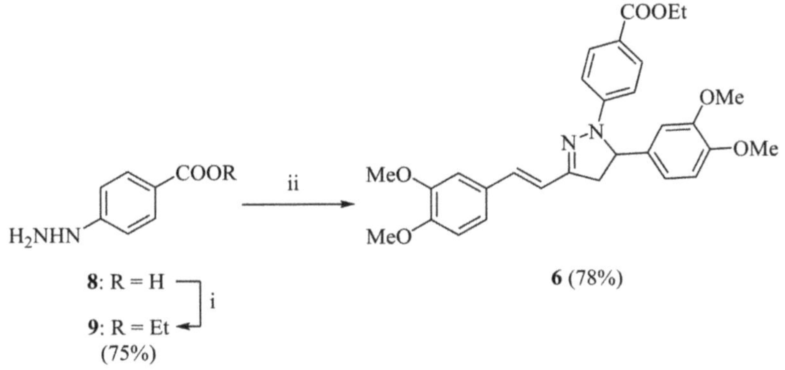

Scheme 2. Synthesis of (*E*)-1-(4-ethoxycarbonylphenyl)-5-(3,4-dimethoxyphenyl)-3-(3,4-dimetho-xystyryl)-2-pyrazoline **6**. Reagents and conditions: (i) H_2SO_4, EtOH, reflux, 5 h; (ii) **7**, glacial AcOH, reflux, 5 h.

In the second step, **7** reacted with ethyl hydrazinobenzoate **9**, prepared from the esterification of the corresponding carboxylic acid **8** with ethanol and concentrated sulfuric acid. The cyclization reaction between curcuminoid **7** and the hydrazine **9** in refluxing glacial acetic acid afforded the desired pyrazoline **6** in high yield and purity without the need for further purification. However, for analytical or biological purposes, even higher purity was achieved by ethanol precipitation ($\geq$ 95%), as determined by high performance liquid chromatography (HPLC, see Supplementary Materials).

The structure of **6** has been confirmed by NMR spectroscopy and HRMS spectrometry. The characteristic signals assigned to the protons on C-4 (H_M, H_A) and C-5 (H_X) of the pyrazoline ring (Figure 1) can be described as a typical AMX spin system. Three doublets of doublets appear at 3.03 (H_A), 3.78 (H_M) and 5.51 (H_X) ppm. The coupling constants conform to this abovementioned pattern having the following J values: J_{AM} = 17.2 Hz the J_{AX} = 4.6 Hz for the *cis*- configuration and J_{MX} = 12.2 Hz for the *trans*. Moreover, the large J value (16.4 Hz) of the vinylic H_α and H_β protons, appearing at 6.81 and 7.21 ppm respectively, is in accordance with the expected *E*-configuration of the double bond. The rest of the ^1H signals, along with the ^{13}C signals, were assigned by means of 2D NMR (COSY, HSQC and HMBC—see Supporting material).

The FTIR spectrum of the prepared compound further supports the formation of a 2-pyrazoline. The characteristic C=N and C–N bonds of the molecule appear at 1600 and 1250 cm^{-1} respectively as strong bands, while the medium intensity band at 1700 cm^{-1} is assigned to the C=O of the aromatic ester.

2.2. Biological Evaluation

2.2.1. DNA Binding Studies

The potential of compound **6** to interact with CT-DNA was studied by means of the biophysical methods of DNA melting and DNA viscosity changes. Firstly, in thermal denaturation experiments for CT-DNA alone, under our experimental conditions, a Tm of 69.02 ± 0.31 °C was recorded, a value that is in good agreement with reported literature values [19]. Upon co-incubation of increasing concentrations of compound **6** with CT-DNA (5×10^{-5} M), a concentration-dependent rise of Tm values was recorded reaching a ΔTm value of 3.53 ± 0.11 °C at the highest [compound **6**]/[DNA] ratio of R = 0.5 (Figure 2A). As in the literature an increase in melting temperature with ΔTm values ranging from +3 °C to +14 °C is associated with intercalative binding [20], the observed ΔTm value suggests that compound **6** may interact with DNA through intercalation. To further investigate the nature of this interaction with CT-DNA, viscosity measurements were carried out. In general, interaction of a compound with DNA results in changes in the length of DNA [21]. When classical intercalation occurs, DNA lengthening is usually observed, due to the separation of base pairs at the intercalation site, which produces a concomitant increase in the relative specific viscosity of such solutions. As it is shown by the viscosity data obtained (Figure 2B), with increasing concentration of compound **6**, the relative viscosity of CT-DNA increases continuously, providing additional evidence of intercalation mode of binding. Curcumin and curcumin derivatives are known to interact with DNA via groove binding [22–25], as demonstrated by UV-visible, competitive fluorescence studies, CD spectroscopy, DNA viscosity and melting studies. The interaction of the pyrazoline curcuminoid **6**, with DNA via the intercalative mode, demonstrates the fact that structural modifications of bioactive molecules modulate dramatically their mechanism, mode and specificity of action.

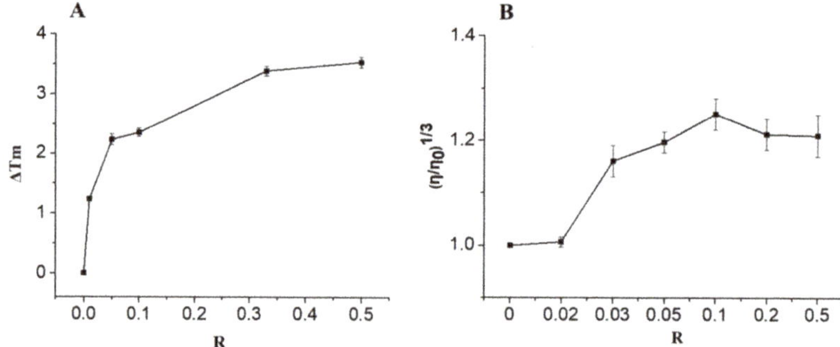

Figure 2. Thermal denaturation data of CT-DNA (**A**) and Effect of increasing amounts on the relative viscosity of CT-DNA (**B**) upon addition of compound **6** with ratios R = [compound **6**]/[DNA] ranging from 0 to 0.5.

2.2.2. In Vitro Cytotoxicity Studies Against Doxorubicin-Resistant Breast Cancer Cells

To assess the relative toxicity of compound **6**, Doxorubicin (DOX) and their combination, the MTT colorimetric method was employed against MCF-7 breast cancer cell line and the corresponding DOX-resistant (MCF-7/DOX^R) cell line. Compound **6** exhibited moderate toxicity against both MCF-7 DOX-sensitive and resistant cells, with IC$_{50}$ values of 53.09 μM and 85.11 μM respectively, following a 48 h treatment (Table 1). Furthermore, the IC$_{50}$ value obtained for the cytotoxity of DOX in MCF-7 was in good agreement with literature data [26], and as expected, in the MCF-7/DOX^R cells DOX was found to be less toxic compared to corresponding non-resistant cells (14-fold lower). In order to evaluate the potential ability of compound **6** to re-sensitize the DOX-resistant cells and re-establish a low IC_{50} value for DOX, or investigate the potential synergistic activity of compound **6** with DOX in the parental cell line, both the MCF-7 and the MCF-7/DOX^R cells were pre-treated with a small amount of compound **6** before the DOX co-incubation. As presented in Table 1, compound **6** did not alter IC_{50} values of DOX in either cell lines and no significant reduction was observed. Our data reveal that even though compound **6** has a close structural similarity with **5** it does not seem to reverse drug resistance in DOX-resistant MCF-7/DOX^R and re-sensitize them as well as no synergistic activity was exhibited with DOX against the parental MCF-7 cells.

Table 1. Effect of curcuminoid pyrazoline **6** on DOX sensitivity in MCF-7 and MCF-7/DOX^R.

Compound	IC_{50} (μM)	
	MCF-7	**MCF-7-DOX^R**
6	53.09 ± 3.58	85.11 ± 4.68
DOX	5.22 ± 1.33	70.90 ± 8.75
DOX (pre-incubated with **6**)	7.71 ± 1.59	72.55 ± 4.99

3. Materials and Methods

3.1. General

All reagents were purchased from Sigma-Aldrich (St. Louis, MO, USA) and Alfa Aesar (Lancaster, UK) unless otherwise stated and used without further purification. All the media/agents for the cultures of cells were purchased from Thermo Fisher Scientific (Cleveland, OH, USA). Melting points were determined with a Gallenkamp MFB-595 melting point apparatus (Weiss Gallenkamp, London, UK). NMR spectra were recorded with a Bruker Avance 500 MHz spectrometer (Bruker, Rheinstetten, Germany) operating at 500 MHz (^{1}H) and 125 (^{13}C). Chemical shifts are reported in ppm relative to DMSO-d_6 (^{1}H: δ = 2.50 ppm, ^{13}C: δ = 39.52 ± 0.06 ppm) or CDCl$_3$ (^{1}H: δ = 7.26 ppm). UV-Vis spectra were recorded with a Hitachi U-3010 spectrophotometer (Hitachi, Tokyo, Japan). IR spectra were recorded on a Perkin-Elmer Spectrum 100-IR spectrophotometer (Perkin-Elmer, San Francisco, CA, USA). HRMS spectra were recorded on UHPLC LC-MSn Orbitrap Velos-Thermo instrument (Thermo Scientific; Bremen, Germany) in the Institute of Biology, Medicinal Chemistry and Biotechnology of the National Hellenic Research Foundation. HPLC analysis was performed with a Waters 600E Chromatography System coupled with a Waters 486 UV detector (at 254 nm) (Waters, Manchester, UK). Separation was achieved on a C-18 reverse phase column (250 × 4 mm, 5μm) eluted with a binary gradient system at 1 mL/min flow rate. Mobile phase A was methanol containing 0.1% trifluoroacetic acid, while mobile phase B was water containing 0.1% trifluoroacetic acid. The elution gradient was 0–1 min 5% A (95% B), followed by a linear gradient to 70% A (30% B) in 5 min; this composition was held for another 3 min; 85% A (15% B) for 17 min. After a column wash with 95% A (5% B) for 10 min, the column was re-equilibrated by applying the initial conditions 5% A (95% B) for 10 min prior to the next injection.

3.2. (1E,4E)-1,5-Bis(3,4-dimethoxyphenyl)penta-1,4-dien-3-one (7)

To a solution of veratraldehyde (3.32 g, 20 mmol) and acetone (0.98 g, 10 mmol) in ethanol (10 mL) was added a suspension of NaOH (5 g, 125 mmol) in ethanol (40 mL). The resulting suspension was stirred for 2 h at room temperature. After 10 min, a yellow solid started to precipitate. The mixture was filtered, washed with water (×3) and dried. The resulting solid was recrystallized from ethanol to afford the product as yellow needles (2.84 g, 7.20 mmol, 72%). Mp: 79–80 °C; ^{1}H NMR (500 MHz, DMSO-d_6): δ 3.82 (s, 6H, OMe), 3.84 (s, 6H, OMe), 7.03 (d, *J* = 8.3 Hz, 2H, Ph), 7.23 (d, *J* = 15.9 Hz, 2H, -C*H*=CH-Ph), 7.33 (d, *J* = 8.3 Hz, 2H, Ph), 7.40 (s, 2H, Ph), 7.70 (d, *J* = 15.9 Hz, 2H, -CH=C*H*-Ph) [27–29].

3.3. Ethyl 4-hydrazinobenzoate (9)

To a solution of 4-hydrazinobenzoic acid (1.52 g, 10 mmol) in absolute ethanol (20 mL) was added dropwise concentrated H_2SO_4 (1 mL). After stirring under reflux for 8 h, the mixture was cooled to room temperature and then concentrated under vacuum. The residue was suspended in AcOEt (40 mL) and washed with a saturated K_2CO_3 solution. The latter was extracted two times with AcOEt (20 mL). The combined organic solutions were dried over Na_2SO_4 and the solvent was evaporated to afford the product as off-white solid (1.35 g, 7.5 mmol, 75%). Mp: 106–107 °C; ^1H NMR (500MHz, CDCl$_3$): δ 1.37 (t, J = 7.0 Hz, 3H, CH_3CH$_2$), 3.31 (br, 2H, NHNH_2), 4.33 (q, J = 7.0 Hz, 2H, CH$_3$CH_2), 5.52 (br, 1H, NHNH$_2$), 7.40 (d, J = 7.4 Hz, 2H, Ph), 7.92 (d, J = 7.4 Hz, 2H, Ph) [30].

3.4. (E)-1-(4-Ethoxycarbonylphenyl)-5-(3,4-dimethoxyphenyl)-3-(3,4-dimethoxystyryl)-2-pyrazoline (6)

A mixture of the curcuminoid **7** (177 mg, 0.50 mmol) and ethyl 4-hydrazinobenzoate **9** (270 mg, 1.50 mmol) was refluxed in glacial acetic acid (6 mL) for 5 h. The solution was cooled to 50 °C and then poured into ice-cold water (50 mL). The completion of the reaction was monitored by TLC (*n*-hexane/AcOEt = 7:3) after mini work-up. The precipitate was filtered and washed thoroughly with water (×3). The product (199 mg, 0.39 mmol, 78%) was dried in a desiccator over $CaCl_2$ overnight. For analytical purposes, it was precipitated from hot ethanol. Yellow solid. Mp: 76-77 °C; UV-Vis (EtOH) λmax (log ε): 388 nm (4.65); FTIR (KBr, cm^{-1}): 1700, 1600, 1510, 1265, 1170, 1105, 1025, 765; ^1H NMR (500 MHz, DMSO-d_6): δ 1.26 (t, J = 6.8 Hz, 3H, CH_3CH$_2$O-), 3.03 (dd, J_{MX} = 4.6 Hz, J_{AM} = 17.1 Hz, 1H, CH$_A$$H_M$), 3.69 (s, 3H, MeO), 3.71 (3H, s, MeO), 3.74 (dd, J_{AX} = 11.8 Hz, J_{AM} = 17.1 Hz, 1H, CH_AH$_M$), 3.77 (s, 3H, MeO), 3.81 (s, 3H, MeO), 4.21 (q, J = 6.8 Hz, 2H, CH$_3$CH_2O-), 5.51 (dd, J_{MX} = 4.6 Hz, 1H, J_{AX} = 11.8 Hz, CH_X), 6.65 (d, J = 7.8 Hz, 1H, Ph), 6.81 (d, J = 16.2 Hz, 1H, H$_\beta$), 6.87 (1H, s, Ph), 6.88 (d, J = 7.8 Hz, 1H, Ph), 6.94 (d, J = 8.4 Hz, 1H, Ph), 6.99 (d, J = 8.6 Hz, 2H, Ph), 7.08 (d, J = 8.4 Hz, 1H, Ph), 7.21 (d, J = 16.2 Hz, 1H, H$_\alpha$), 7.26 (s, 1H, Ph), 7.74 (d, J = 8.6 Hz, 2H, Ph); ^{13}C NMR (125 MHz, DMSO-d_6): δ 14.3 (CH$_3$CH$_2$), 42.0 (C-4), 55.46 (2C, CH$_3$O-), 55.48 (2C, CH$_3$O-), 59.8 (CH$_3$CH$_2$), 61.8 (C-5), 109.2 (CH-arom), 109.5 (CH-arom), 111.7 (CH-arom), 112.0 (2C, CH-arom), 112.2 (CH-arom), 117.3 (CH-arom), 118.8 (C$_\alpha$), 120.8 (CH-arom), 129.2 (2C, Cq-arom), 130.6 (2C, CH-arom), 134.0 (Cq-arom), 135.1 (C$_\beta$), 146.9 (Cq-arom), 148.1 (Cq-arom), 149.0 (Cq-arom), 149.1 (Cq-arom), 149.4 (Cq-arom), 151.9 (C-3), 165.6 (-COOEt); HRMS: calcd for $C_{30}H_{33}N_2O_6$ (M$^+$ + H) 517.2333; found 517.2328; HPLC: t_R = 16.5 min.

3.5. Preparation of DNA Samples

All DNA experiments were performed in phosphate buffer (0.05 M, pH = 7.2) consisting of Na_2HPO_4 and KH_2PO_4. The CT-DNA solution was sufficiently free of protein as evidenced by the ratio of its UV absorbance at 260 and 280 nm that was approx. 1.9:1 [20]. The DNA concentration per nucleotide was determined spectrophotometrically by using the molar absorption coefficient of ε = 6600 M^{-1}·cm^{-1} per nucleotide at 260 nm. Before each measurement, fresh stock solutions of CT-DNA and compound **6** (in DMSO at a concentration of 10^{-3} M) were prepared and the corresponding mixtures were incubated for 24 h at 25 °C to reach the equilibrium state. Each reported measurement value is the average of three independent experiments.

3.6. Thermal Denaturation Studies

DNA melting temperature measurements were performed using a Varian Cary 300 spectrophotometer (Agilent Technologies, Inc., Santa Clara, CA, USA) equipped with a heating multiple cell block apparatus. Temperatures were maintained under computer control and were increased at a 0.5 °C/min rate. DNA melting experiments were carried out by monitoring the absorbance of DNA at 258 nm in the temperature range from 25.0 °C to 95.0 °C. Solutions were allowed to equilibrate for 1 min at each temperature. Cuvettes (1.0 cm capped quartz) were mounted in a thermal block and the solution temperatures were monitored by a thermistor in the reference cuvette. The melting temperature (Tm) of DNA was determined as the midpoint of the optically detected transition and the measurement was repeated three times for each sample. The melting experiments were performed in phosphate buffer solution (final percentage of DMSO = 2%) by keeping the DNA concentration constant (5×10^{-5} M) while varying the concentration of compound **6** ($0 - 2.5 \times 10^{-5}$ M) to achieve ratios R = [compound **6**]/[DNA] of 0, 0.01, 0.05, 0.1, 0.33, 0.5.

3.7. Viscosity Studies

Viscosity measurements were carried out using Ostwald's viscometer (Schott geräte, type 531 01/0a, thermostat Schott geräte GT 1150, meter Schott geräte AVS 300) at 25 °C (Schott –Instruments GmbH, Germany). The viscometer was thermostated at 25.0 ± 0.1 °C in a constant temperature bath. Flow time of solutions was measured with a digital stopwatch and each sample was measured three times and an average flow time was calculated. The values did not differ by more than 0.2 s from each other. All solutions were filtered through 0.22 μm filter (Millipore, Billerica, MA) prior to the measurements. Experiments were carried out in phosphate buffer solution (final percentage of DMSO = 2%) by keeping the DNA concentration constant (5×10^{-5} M) while varying the concentration of compound **6** ($0–2.5 \times 10^{-5}$ M) to achieve ratios R = [compound **6**]/[DNA] of 0, 0.02, 0.033, 0.05, 0.1, 0.2, 0.5. The intrinsic viscosity [n] was calculated according to the relation [n] = $(t - t_b)/t_b$, where t_b is the flow time for the buffer and t is the observed flow time for DNA in the presence or absence of the compound **6**. Data were presented as $(n/n_0)^{1/3}$ versus R where n and n_0 are the intrinsic viscosities in the presence or absence of the compound **6**. For the low DNA concentrations used in these experiments, the intrinsic viscosity [n] is proportional to the difference in the flow time for the buffer with and without DNA, resulting in the following Equation (1) [31]:

$$\frac{L}{L_0} = \frac{[n]^{1/3}}{[n_0]^{1/3}} = \frac{(t - t_b)^{1/3}}{(t_0 - t_b)^{1/3}} \tag{1}$$

where L and L_0 are the DNA lengths and [n] and [n_0] are the intrinsic viscosities with and without the compound. The t_b, t_0 and t are the flow times of the buffer, the plain DNA and the DNA/compound solution, respectively.

3.8. Cell Culture

Human breast adenocarcinoma MCF-7 cells (American Type Culture Collection, Manassas, VA, USA), were grown in DMEM growth medium of pH 7.4 supplemented with 10% FBS, 100 U/mL of penicillin, 2 mM glutamine and 100 μg/mL of streptomycin. Cell cultures were maintained in flasks and were grown at 37 °C in a humidified atmosphere of 5% CO_2 in air. Subconfluent cells were detached using a 0.25% (*w/v*) trypsin—0.03% (*w/v*) EDTA solution and the split ratio was 1:3–1:5. Doxorubicin-resistance was established by stepwise exposure of MCF-7 cells to increased concentrations of DOX ranging between 0.01 and 1 μg/mL. Finally, the MCF-7/DOX[R] cells were able to grow continually in medium with 0.1 μg/mL DOX. The stock solutions of compound **6** and DOX at a concentration of 10 mM dissolved in DMSO were diluted to different concentrations as required.

3.9. In Vitro Cytotoxicity Assay

In vitro cytotoxicity of compound **6**, DOX and the combination of them against the MCF-7 and MCF-7/DOX[R] cell lines was determined by the MTT colorimetric assay. Cells were seeded in 96-well plates (10^4 cells/well) and grown overnight at 37 °C in a 5% CO_2 incubator. Exponentially growing cells were incubated for 48 h with various concentrations ranging between $10^{-3}–10^{-8}$ M of compound **6** and DOX and the final DMSO concentration never exceeded 0.2%. For the combination experiments, MCF-7/DOX[R] cells were pre-incubated for 1 h with compound **6** (10^{-5} M) and then DOX was added in various concentrations ($10^{-3}–10^{-8}$ M) for another 47 h. The medium was then removed and replaced with 100 μL of MTT solution (Sigma-Aldrich, Darmstadt, Germany, 1 mg mL^{-1}). After a 4 h incubation, the solution was aspirated, formazan crystals were solubilized in 100 µL of dimethyl sulfoxide (DMSO) and absorbance was recorded at 540 nm (Tecan well plate reader, Tecan, Grödig, Austria). The results were expressed as % cell viability = (mean optical density (OD) of treated cells/mean OD of untreated cells) × 100. Data were plotted against the corresponding compound concentration in a semi-log chart and the values of IC_{50} (the concentration of test compound required to reduce the fraction of live cells to 50% of the control) were calculated from the dose–response curves using GraphPad Prism 5.0 software (GraphPad Software Inc., CA, USA).

4. Conclusions

(*E*)-1-(4-Ethoxycarbonylphenyl)-5-(3,4-dimethoxyphenyl)-3-(3,4-dimethoxystyryl)-2-pyrazoline was successfully synthesized for the first time from the corresponding monocarbonyl acetone-derived curcuminoid and ethyl hydrazinobenzoate in high yield and purity. The compound exhibited significant intercalation binding activity with CT-DNA, which renders it a suitable lead compound for further derivatisation aiming to improve its DNA binding capacity. However, despite its structural similarity with the corresponding curcumin pyrazoline derivative which has shown MDR reversal activity no re-sensitisation of DOX-resistant breast cancer was observed and there was no synergistic activity with DOX against non-resistant cell lines. Further investigation is underway to identify whether conversion of the pyrazoline to pyrazole moiety may improve the MDR reversal activity of such monocarbonyl curcumin derivatives.

Supplementary Materials: The following are available online. All the spectroscopic data of the title compound **6** namely ^{1}H and ^{13}C NMR, ^{1}H-^{1}H COSY NMR, HSQC and HMBC NMR, FT-IR, HRMS, UV-Vis spectra, and HPLC are available online.

Author Contributions: Conceptualization, M.S.; methodology and data analysis, D.M and B.M.; 2D NMR analysis A.P. and M.P.; writing—original draft preparation, D.M., B.M., M.S.; writing—review and editing and funding acquisition, M.P.; supervision, M.S., C.M. and M.P. All authors have read and agreed to the published version of the manuscript.

Funding: The authors acknowledge financial support of this work by the project "Target Identification and Development of Novel Approaches for Health and Environmental Applications" (MIS 5002514) which is implemented under the Action for the Strategic Development on the Research and Technological Sectors, funded by the Operational Programme "Competitiveness, Entrepreneurship and Innovation" (NSRF 2014-2020) and co-financed by Greece and the European Union (European Regional Development Fund). B. Mavroidi gratefully acknowledges financial support by Stavros Niarchos Foundation (SNF) through implementation of the program of Industrial Fellowships at NCSR "Demokritos".

Conflicts of Interest: The authors declare no conflict of interest. The funders had no role in the design of the study; in the collection, analyses, or interpretation of data; in the writing of the manuscript, or in the decision to publish the results.

References

1. *Natural Products in Medicinal Chemistry*; Hanessian, S., Ed.; Wiley-VCH:: Weinheim, Germany, 2014.
2. Vitaku, E.; Smith, D.T.; Njardarson, J.T. Analysis of the structural diversity, substitution patterns, and frequency of nitrogen heterocycles among U.S. FDA approved pharmaceuticals. *J. Med. Chem.* **2014**, *57*, 10257–10274. [CrossRef] [PubMed]
3. Varghese, B.; Al-Busafi, S.N.; Suliman, F.O.; Al-Kindy, M.Z. Unveiling a versatile heterocycle: Pyrazoline—A review. *RSC Adv.* **2017**, *7*, 46999–47016. [CrossRef]
4. Shaaban, M.R.; Mayhoub, A.S.; Farag, A.M. Recent advances in the therapeutic applications of pyrazolines. *Expert Opim. Ther. Patents* **2012**, *22*, 253–291. [CrossRef] [PubMed]
5. Akhtar, J.; Khan, A.A.; Ali, Z.; Haider, R.; Yar, M.S. Structure-activity relationship (SAR) study and design strategies of nitrogen-containing heterocyclic moieties for their anticancer activities. *Eur. J. Med. Chem.* **2017**, *125*, 143–189. [CrossRef] [PubMed]
6. Havrylyuk, D.; Zimenkovsky, B.; Vasylenko, O.; Zaprutko, L.; Gzella, A.; Lesyk, R. Synthesis of novel thiazolone-based compounds containing pyrazoline moiety and evaluation of their anticancer activity. *Eur. J. Med. Chem.* **2009**, *44*, 1396–1404. [CrossRef] [PubMed]

7. Shaharyar, M.; Abdullah, M.M.; Bakht, M.A.; Majeed, J. Pyrazoline bearing benzimidazoles: Search for anticancer agent. *Eur. J. Med. Chem.* **2010**, *45*, 114–119. [CrossRef]

8. Insuasty, B.; Montoya, A.; Becerra, D.; Quiroga, J.; Abonia, R.; Robledo, S.; Vélez, I.D.; Upegui, Y.; Nogueras, M.; Cobo, J. Synthesis of novel analogs of 2-pyrazoline obtained from [(7-chloroquinolin-4-yl)amino]chalcones and hydrazine as potential antitumor and antimalarial agents. *Eur. J. Med. Chem.* **2013**, *67*, 252–262. [CrossRef]

9. Shin, S.Y.; Yoon, H.; Hwang, D.; Ahn, S.; Kim, D.-W.; Koh, D.; Lee, Y.H.; Lim, Y. Benzochalcones bearing pyrazoline moieties show anti-colorectal cancer activities and selective inhibitory effects on aurora kinases. *Bioorg. Med. Chem.* **2013**, *21*, 7018–7024. [CrossRef]

10. Rathish, I.G.; Javed, K.; Ahmad, S.; Bano, S.; Alam, M.S.; Pillai, K.K.; Singh, S.; Bagchi, V. Synthesis and antiinflammatory activity of some new 1,3,5-trisubstituted pyrazolines bearing benzene sulfonamide. *Bioorg. Med. Chem. Lett.* **2009**, *19*, 255–258. [CrossRef]

11. Bansal, E.; Srivastava, V.K.; Kumar, A. Synthesis and anti-inflammatory activity of 1-acetyl-5-substitute daryl-3-(b-aminonaphthyl)-2-pyrazolines and b-(substituted aminoethyl) amidonaphthalenes. *Eur. J. Med. Chem.* **2001**, *36*, 81–92. [CrossRef]

12. Lokeshwari, D.M.; Achutha, D.K.; Srinivasan, B.; Shivalingegowda, N.; Krishnappagowda, L.N.; Kariyappa, A.K. Synthesis of novel 2-pyrazoline analogues with potent anti-inflammatory effect mediated by inhibition of phospholipase A2: Crystallographic, in silico docking and QSAR analysis. *Bioorg. Med. Chem. Lett.* **2017**, *27*, 3806–3811. [CrossRef] [PubMed]

13. Karthikeyan, M.S.; Holla, B.S.; Kumari, N.S. Synthesis and antimicrobial studies on novel chloro-fluorine containing hydroxy pyrazolines. *Eur. J. Med. Chem.* **2007**, *42*, 30–36. [CrossRef] [PubMed]

14. Siddiqui, Z.N.; Musthafa, T.N.M.; Ahmad, A.; Khan, A.U. Thermal solvent-free synthesis of novel pyrazolyl chalcones and pyrazolines as potential antimicrobial agents. *Bioorg. Med. Chem. Lett.* **2011**, *21*, 2860–2865. [CrossRef] [PubMed]

15. Mishra, V.K.; Mishra, M.; Kashaw, V.; Kashaw, S.K. Synthesis of 1,3,5-trisubstituted pyrazolines as potential antimalarial and antimicrobial agents. *Bioorg. Med. Chem.* **2017**, *25*, 1949–1962. [CrossRef] [PubMed]

16. Manna, F.; Chimenti, F.; Fioravanti, R.; Bolasco, A.; Secci, D.; Chimenti, P.; Ferlini, C.; Scambia, G. Synthesis of some pyrazole derivatives and preliminary investigation of their affinity binding to P-glycoprotein. *Bioorg. Med. Chem. Lett.* **2005**, *15*, 4632–4635. [CrossRef]

17. Kolotova, E.S.; Shtil, A.A.; Novikov, F.N.; Chilov, G.G.; Stroganov, O.V.; Stroilov, V.S.; Zeifman, A.A.; Titov, I.Y.; Sagnou, M.; Alexiou, P. Novel Derivatives of 3,5-Divinyl-pyrazole for Medical Application. WO 2016/190770 A1, 1 December 2016.

18. Gottesman, M.M.; Pastan, I.; Ambudkar, S.V. P-Glycoprotein and multidrug resistance. *Curr. Opin. Gen. Dev.* **1996**, *6*, 610–617. [CrossRef]

19. Halevas, E.; Mavroidi, B.; Swanson, C.H.; Smith, G.C.; Moschona, A.; Hadjispyrou, S.; Salifoglou, A.; Pantazaki, A.A.; Pelecanou, M.; Litsardakis, G. Magnetic cationic liposomal nanocarriers for the efficient drug delivery of a curcumin-based vanadium complex with anticancer potential. *J. Inorg. Biochem.* **2019**, *199*, 110778. [CrossRef]

20. Mavroidi, B.; Sagnou, M.; Stamatakis, K.; Paravatou-Petsotas, M.; Pelecanou, M.; Methenitis, G. Palladium(II) and platinum(II) complexes of derivatives of 2-(4′-aminophenyl)benzothiazole as potential anticancer agents. *Inorg Chim. Acta* **2016**, *444*, 63–75. [CrossRef]

21. Suh, D.; Chaires, J.B. Criteria for the mode of binding of DNA binding agents. *Bioorg. Med. Chem.* **1995**, *3*, 723–728. [CrossRef]

22. Zsila, F.; Bikádi, Z.; Simonyi, M. Circular dichroism spectroscopic studies reveal pH dependent binding of curcumin in the minor groove of natural and synthetic nucleic acids. *Org. Biomol. Chem.* **2004**, *2*, 2902–2910.

23. Kunwar, A.; Simon, E.; Singh, U.; Chittela, R.; Sharma, D.; Sandur, S.K.; Priyadarsini, I.K. Interaction of a curcumin analogue dimethoxycurcumin with DNA. *Chem. Biol. Drug Des.* **2011**, *77*, 281–287. [CrossRef] [PubMed]

24. Sahoo, B.K.; Ghosh, K.S.; Bera, R.; Dasgupta, S. Studies on the interaction of diacetylcurcumin with calf thymus-DNA. *Chem. Phys.* **2008**, *351*, 163–169. [CrossRef]

25. Bera, R.; Sahoo, B.K.; Ghosh, K.S.; Dasgupta, S. Studies on the interaction of isoxazolcurcumin with calf thymus DNA. *Int. J. Biol. Macromol.* **2008**, *42*, 14–21.

26. Arif, I.S.; Hooper, C.L.; Greco, F.; Williams, A.C.; Boateng, S.Y. Increasing doxorubicin activity against breast cancer cells using PPARγ-ligands and by exploiting circadian rhythms. *Br. J. Pharmacol.* **2013**, *169*, 1178–1188. [CrossRef]

27. Li, Q.; Chen, J.; Luo, S.; Xu, J.; Huang, Q.; Liu, T. Synthesis and assessment of the antioxidant and antitumor properties of asymmetric curcumin analogues. *Eur. J. Med. Chem.* **2015**, *93*, 461–469. [CrossRef]

28. Butcher, R.J.; Jasinski, J.P.; Yathirajan, H.S.; Bindya, S.; Narayana, B.; Sarojini, B.K. 1,5-Bis(3,4-dimethoxyphenyl)penta-1,4-dien-3-one. *Acta Cryst.* **2007**, *E63*, o3115. [CrossRef]

29. Du, Z.-Y.; Bao, Y.-D.; Liu, Z.; Qiao, W.; Ma, L.; Huang, Z.-S.; Gu, L.-Q.; Chan, A.S.C. Curcumin analogs as potent aldose reductase inhibitors. *Arch. Pharm. Chem. Life Sci.* **2006**, *339*, 123–128. [CrossRef]

30. Bourrie, B.; Casellas, P.; Ciapetti, P.; Derocq, J.-M.; Jegham, S.; Muneaux, Y.; Wermuth, C.-G. Pyridoindolone Derivatives Substituted in the 3-Position by a Phenyl, Their Preparation and Their Application in Therapeutics. US 7,390,818 B2, 24 June 2008.

31. Cohen, G. Eisenber, Viscosity and sedimentation study of sonicated DNA–proflavine complexes. *Biopolymers* **1969**, *8*, 45. [CrossRef]

Short Note

Ethyl 11a,12-Dihydrobenzo[*b*]benzo[5,6][1,4]oxazino[2,3-*e*][1,4]oxazine-5a(6*H*)-carboxylate

Lidia S. Konstantinova [1,2], **Mikhail A. Tolmachev** [1], **Vadim V. Popov** [2] and **Oleg A. Rakitin** [1,2,*]

[1] N. D. Zelinsky Institute of Organic Chemistry Russian Academy of Sciences, 47 Leninsky Prospekt, 119991 Moscow, Russia; konstantinova_ls@mail.ru (L.S.K.); mtolmachev4@gmail.com (M.A.T.)

[2] Nanotechnology Education and Research Center, South Ural State University, 76 Lenina Avenue, 454080 Chelyabinsk, Russia; popov.ioc@gmail.com

* Correspondence: orakitin@ioc.ac.ru; Tel.: +7-499-1355327

Received: 30 June 2020; Accepted: 18 July 2020; Published: 21 July 2020

Abstract: The 11a,12-dihydrobenzo[*b*]benzo[5,6][1,4]oxazino[2,3-*e*][1,4]oxazine heterocyclic system has been used in the construction of heteropropellanes, which attracted much attention not only on the possible modification of drugs, but also for novel materials with unusual and important physical properties. In this communication, the reaction of ethyl 2-(hydroxyimino)propanoate **1** with disulfur dichloride and o-aminophenol, which gave ethyl 11a,12-dihydrobenzo[*b*]benzo[5,6][1,4]oxazino[2,3-*e*][1,4]oxazine-5a(6*H*)-carboxylate in moderate yield, was described. The structure of the newly synthesized compound was established by means of elemental analysis, high resolution mass-spectrometry, ^{1}H, ^{13}C NMR and IR spectroscopy, mass-spectrometry and X-ray analysis.

Keywords: oxygen-nitrogen heterocycles; 11a,12-dihydrobenzo[*b*]benzo[5,6][1,4]oxazino[2,3-*e*] [1,4]oxazine; disulfur dichloride; o-aminophenol; condensation

---

1. Introduction

Disulfur dichloride is a well-known and frequently used sulfurating agent [1,2]. Sometimes it may act as oxidizing agent as well, giving compounds without additional sulfur atoms [3]. It was discovered that by the reaction of substituted acetoximes, including ethyl 2-(hydroxyimino)propanoate, with S_2Cl_2 4-substituted 1,2,3-dithiazolium salts are formed, which by treatment with oxygen, sulfur and nitrogen nucleophiles afforded 1,2,3-dithiazole 5-ones, 5-thiones and 5-phenylimines, respectively [4]. Recently it was found that, when o-aminophenol was added in the final step, 2-((4-aryl(hetaryl)-5*H*-1,2,3-dithiazol-5-ylidene)amino)phenols were isolated, although in low yield [5]. Herein, we report the synthesis of previously unknown ethyl 11a,12-dihydrobenzo[*b*]benzo[5,6][1,4]oxazino[2,3-*e*][1,4]oxazine-5a(6*H*)-carboxylate, which unexpectedly formed in the reaction of ethyl 2-(hydroxyimino)propanoate with disulfur dichloride and o-aminophenol.

2. Results and Discussion

Continuing the study of ethyl 2-(hydroxyimino)propanoate reactivity, it was treated with S_2Cl_2, pyridine in MeCN followed by addition of o-aminophenol. Unexpectedly, a new compound, **1**, a yellow solid $C_{17}H_{16}N_2O_4$, was being formed instead of the expected ethyl 5-((2-hydroxyphenyl)imino)-5*H*-1,2,3-dithiazole-4-carboxylate, **2**. According to NMR, mass and elemental analysis data compound **1** is formally a product in which two aminophenol molecules are cross-linked by two carbon atoms of ethyl 2-(hydroxyimino)propanoate. Structure **1** was finally proven by X-ray analysis as ethyl 11a,12-dihydrobenzo[*b*]benzo[5,6][1,4]oxazino[2,3-*e*][1,4]oxazine-5a(6*H*)-carboxylate (Scheme 1).

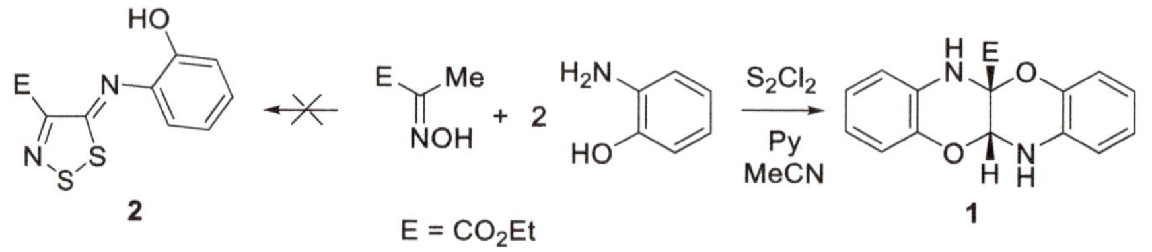

Scheme 1. Synthesis of ethyl 11a,12-dihydrobenzo[*b*]benzo[5,6][1,4]oxazino[2,3-*e*][1,4]oxazine-5a(6*H*)-carboxylate **1**.

It was assumed that ethyl 2-(hydroxyimino)propanoate may be dezoximated by S_2Cl_2 to ethyl 2-oxopropanoate and then reacted with sulfur containing species, such as S_2Cl_2 or S_8, formed in this reaction, and *o*-aminophenol to give benzoxazinobenzoxazine **1**. Similar transformation was described for 1,2-diphenylethan-1-one oxime (PhCH$_2$C(O)Ph) with the formation of 5a,11a-diphenyl-5a,6,11a,12-tetrahydrobenzo[*b*]benzo[5,6][1,4]oxazino[2,3-*e*][1,4]oxazine **3** (Scheme 2) [6].

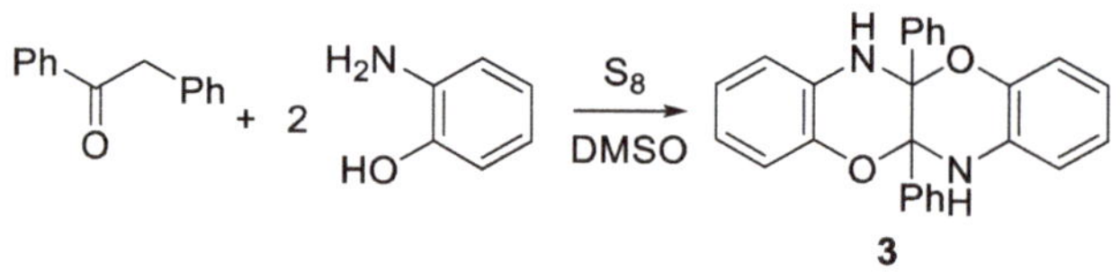

Scheme 2. Synthesis of 5a,11a-diphenyl-5a,6,11a,12-tetrahydrobenzo[*b*]benzo[5,6][1,4]oxazino-[2,3-*e*][1,4]oxazine **3** from 1,2-diphenylethan-1-one oxime.

We checked this possibility and found that ethyl 2-oxopropanoate did not react with S_2Cl_2 and *o*-aminophenol. Therefore, the role of ethyl 2-(hydroxyimino)propanoate is crucial for the formation of oxazinooxazine **1**, and it is necessary to find another mechanistic explanation for this reaction. So, the mechanism of this transformation is still unclear and requires further investigation.

The structure of oxazinooxazine **1** was confirmed by means of elemental analysis, high resolution mass-spectrometry, ^{1}H, ^{13}C NMR and IR spectroscopy, mass-spectrometry and X-ray analysis (Figure 1) (see Supplementary Materials).

In conclusion, the synthesis of previously unknown ethyl 11a,12-dihydrobenzo[b]benzo[5,6][1,4]oxazino[2,3-*e*][1,4]oxazine-5a(6*H*)-carboxylate **1** from ethyl 2-(hydroxyimino)propanoate and disulfur dichloride was developed. The described experimental procedure may serve as an efficient basis for the synthesis of other fused oxazinooxazines. Fused with benzene rings, oxazinooxazines are important heterocyclic scaffold in the construction of heteropropellanes: structurally interesting propeller-like molecules with application in material sciences and medicinal chemistry [7].

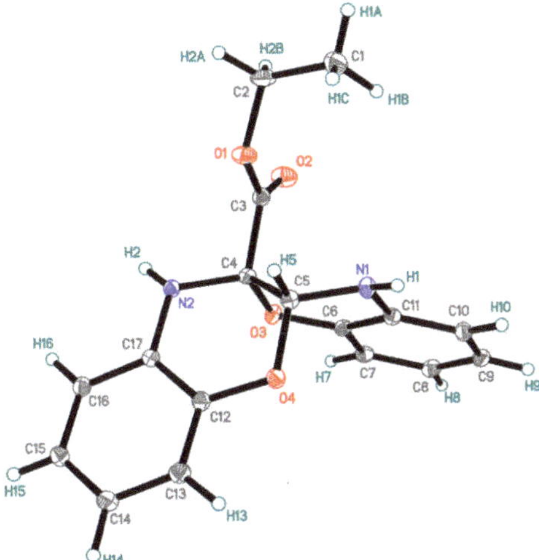

Figure 1. X-ray structure of ethyl 11a,12-dihydrobenzo[*b*]benzo[5,6][1,4]oxazino-[2,3-*e*][1,4]oxazine-5a(6*H*)-carboxylate **1**. Thermal ellipsoids are at 50% probability.

3. Experimental Section

3.1. General Information

The solvents and reagents were purchased from commercial sources and used as received. Elemental analysis was performed on a 2400 Elemental Analyzer (Perkin Elmer Inc., Waltham, MA, USA). Melting point was determined on a Kofler hot-stage apparatus and is uncorrected. ^{1}H were taken with a Bruker AM-300 machine (Bruker AXS Handheld Inc., Kennewick, WA, USA) (at frequencies of 300.1) and ^{13}C NMR spectra were taken with a Bruker DRX-500 machine (Bruker AXS Handheld Inc., Kennewick, WA, USA) (125.8 MHz) in DMSO-d_6 solution, with TMS as the standard. J values are given in Hz. MS spectrum (EI, 70 eV) was obtained with a Finnigan MAT INCOS 50 instrument (Hazlet, NJ, USA). IR spectrum was measured with a Bruker "Alpha-T" instrument in KBr pellet. High-resolution MS spectrum was measured on a Bruker micrOTOF II instrument (Bruker Daltonik Gmbh, Bremen, Germany) using electrospray ionization (ESI). The measurement was performed in a positive ion mode (interface capillary voltage: 4500 V) or in a negative ion mode (3200 V); mass range was from *m/z* 50 to *m/z* 3000 Da; external or internal calibration was performed with Electrospray Calibrant Solution (Fluka). Syringe injection was used for solutions in acetonitrile, methanol, or water (flow rate 3 L/min^{-1}). Nitrogen was applied as a dry gas; interface temperature was set at 180 °C.

Crystal structure determination was performed in the Department of Structural Studies of Zelinsky Institute of Organic Chemistry, Moscow. X-ray diffraction data were collected at 100 K on a Bruker Quest D8 diffractometer (Bruker Corporation, Germany) equipped with a Photon-III area-detector (graphite monochromator, shutterless φ- and ω-scan technique), using Mo K$_\alpha$-radiation (0.71073 Å). The intensity data were integrated by the SAINT program and were corrected for absorption and decay using SADABS [8]. The structure was solved by direct methods using SHELXS-2013 and refined on F^2 using SHELXL-2018. All non-hydrogen atoms were refined with individual anisotropic displacement parameters. The positions of all hydrogen atoms were found from the electron density-difference map; these atoms were refined with individual isotropic displacement parameters. Cambridge Crystallographic Data Centre contains the supplementary crystallographic data for this paper No. CCDC 2012896. These data can be obtained free of charge via http://www.ccdc.cam.ac.uk/conts/retrieving.html (or from the CCDC, 12 Union Road, Cambridge CB2 1EZ, UK; Fax: +44 1223 336033; E-mail: deposit@ccdc.cam.ac.uk).

3.2. Synthesis of Ethyl 11a,12-Dihydrobenzo[b]benzo[5,6][1,4]oxazino[2,3-e][1,4]oxazine-5a(6H)-carboxylate 1

Pyridine (0.24 mL, 3 mmol) was added dropwise at 0–5 °C to a stirred solution of ethyl 2-(hydroxyimino)propanoate (1 mmol) and sulfur monochloride (0.16 mL, 2 mmol) in acetonitrile (10 mL) under inert atmosphere of argon. The mixture was stirred at 0 °C for 15–40 min. Then *o*-aminophenol (109 mg, 1 mmol) was added, the mixture was stirred at 0 °C for 30 min, and followed by addition of pyridine (0.16 mL, 2 mmol). The reaction mixture was stirred at room temperature for 2 h; filtered and solvents were evaporated. The residue was separated by column chromatography (Silica gel Merck 60, light petroleum and then light petroleum–CH_2Cl_2 mixtures). Yield 32 mg (11.2%), white crystals, mp 189–191 °C. R_f = 0.56 (CH_2Cl_2). IR (KBr), ν, cm^{-1}: 3405 (N-H), 3360 (N-H), 3058, 2968, 2929, 2861, 1892, 1743, 1601, 1500, 1431, 1305, 1285, 1261, 1228, 1116, 1113, 1051, 998, 967, 931, 884, 832, 782, 750, 703, 611, 578, 537, 507, 460. ^{1}H NMR (ppm, *J*/Hz): δ 6.94–6.75 (m, 8H, Ar), 5.51 (br s, 1H), 5.12 (d, 2H, NH, *J* = 9.17), 4.30–4.23 (m, 2H, CH_2), 1.27 (t, 3H, CH_3, *J* = 7.15). ^{13}C NMR (ppm): δ 166.8, 141.9, 141.4, 128.7, 128.2, 122.6, 122.5, 121.3, 121.1, 117.2, 116.9, 115.6, 115.4, 110.4, 76.5, 63.0, 14.0. MS (EI, 70 Ev), *m/z* (I, %): 312 (M^+, 100), 286 (10), 239 (24), 214 (13), 205 (76), 191 (11), 168 (20), 159 (11), 144 (12), 131 (19), 120 (61), 97 (9). HRMS (ESI-TOF): calcd for $C_{17}H_{16}N_2O_4$ $[M + H]^+$ 313.1183; found *m/z* 313.1186. Anal. calcd. for $C_{17}H_{16}N_2O_4$: C, 65.38; H, 5.16; N, 8.97; found: C, 65.24; H, 5.02; N, 9.16%.

Crystallographic data are given in Table 1.

Table 1. Crystal data and structure refinement for compound **1**.

Empirical Formula	$C_{17}H_{16}N_2O_4$
Formula weight	312.32
Temperature	100(2) K
Wavelength	0.71073 Å
Crystal system	Orthorhombic
Space group	$P2_12_12_1$
Unit cell dimensions	a = 5.5972(2) Å
	b = 15.0183(6) Å
	c = 16.7717(6) Å
Volume	1409.84(9) $Å^3$
Z	4
Density (calculated)	1.471 g/cm^3
Absorption coefficient	0.106 mm^{-1}
F(000)	656
Crystal size	$0.59 \times 0.10 \times 0.09$ mm^3
Theta range for data collection	2.712 to 33.155°
Reflections collected	34967
Independent reflections	5372 [R(int) = 0.0477]
Completeness to theta = 25.242°	99.8%
Absorption correction	Semi-empirical from equivalents
Max. and min. transmission	0.8499 and 0.8299
Refinement method	Full-matrix least-squares on F^2
Data/restraints/parameters	5372/0/272
Goodness-of-fit on F^2	1.040
Final R indices [I > 2sigma(I)]	R1 = 0.0345, wR2 = 0.0855
R indices (all data)	R1 = 0.0423, wR2 = 0.0914
Absolute structure parameter	−0.2(3)
Largest diff. peak and hole	0.384 and −0.201 $e·Å^{-3}$

Supplementary Materials: The following are available online. CIF file, copies of ^{1}H, ^{13}C NMR, IR and mass-spectra for the compound **1**.

Author Contributions: M.A.T. and V.V.P. synthetic experiments; L.S.K. analysis of experimental results and NMR data, O.A.R. writing the paper, supervision and project administration. All authors have read and agreed to the published version of the manuscript.

Funding: The authors are grateful to the Ministry of Education and Science of the Russian Federation (FENU-2020-0019 (2020073GZ)) for financial support.

Conflicts of Interest: The authors declare no conflict of interest.

References

1. Konstantinova, L.S.; Rakitin, O.A. Sulfur monochloride in organic synthesis. *Russ. Chem. Rev.* **2014**, *83*, 225–250. [CrossRef]

2. Konstantinova, L.S.; Rakitin, O.A. Design of sulfur heterocycles with sulfur monochloride: Retrosynthetic analysis and prospects. *Mendeleev Commun.* **2009**, *19*, 55–61. [CrossRef]

3. Konstantinova, L.S.; Rakitin, O.A. Sulfur monochloride in the synthesis of heterocyclic compounds. *Adv. Heterocycl. Chem.* **2008**, *96*, 175–229. [CrossRef]

4. Konstantinova, L.S.; Bol'shakov, O.I.; Obruchnikova, N.V.; Laborie, H.; Tonga, A.; Sopéna, V.; Lanneluc, I.; Picot, L.; Sablé, s.; Thiéry, V.; et al. One-pot synthesis of 5-phenylimino, 5-thione and 5-one-1,2,3-dithiazoles and evaluation of their antimicrobial and antitumor activity. *Bioorg. Med. Chem. Lett.* **2009**, *19*, 136–141. [CrossRef] [PubMed]

5. Baranovsky, I.V.; Konstantinova, L.S.; Tolmachev, M.A.; Popov, V.V.; Lyssenko, K.A.; Rakitin, O.A. *Synthesis of 2-((2-(benzo[d]oxazol-2-yl)-2H-imidazol-4-yl)amino)-phenols from 2-((5H-1,2,3-dithiazol-5-ylidene)amino)phenols through Unprecedented Formation of Imidazole Ring from Two Methanimino Groups*, N. D. Zelinsky Institute of Organic Chemistry Russian Academy of Sciences: Moscow, Russia, 2020, unpublished work.

6. Nguyen, T.B.; Retailleau, P. Sulfur-catalyzed stereo and regioselective synthesis of heteropropellanes via oxidative condensation of cyclohexanones with 2-aminophenols. *Adv. Synth. Catal.* **2019**, *361*, 3588–3592. [CrossRef]

7. Dilmaç, M.; Spuling, E.; de Meijere, A.; Bräse, S. Propellanes—from a chemical curiosity to "explosive" materials and natural products. *Angew. Chem. Int. Ed.* **2017**, *56*, 5684–5718. [CrossRef] [PubMed]

8. Krause, L.; Herbst-Irmer, R.; Sheldrick, G.M.; Stalke, D. Comparison of silver and molybdenum microfocus X-ray sources for single-crystal structure determination. *J. Appl. Cryst.* **2015**, *48*, 3–10. [CrossRef] [PubMed]

MDPI

Short Note

(*E*)-3-((2-Fluorophenyl)(hydroxy)methylene)imidazo[1,2-*a*]pyridin-2(3*H*)-one

Giammarco Tenti, Ángel Cores, María Teresa Ramos and J. Carlos Menéndez *

Unidad de Química Orgánica y Farmacéutica, Departamento de Química en Ciencias Farmacéuticas, Facultad de Farmacia, Universidad Complutense, 28040 Madrid, Spain; giammarco.tenti@hotmail.it (G.T.); acores@ucm.es (Á.C.); mtramos@ucm.es (M.T.R.)
* Correspondence: josecm@ucm.es; Tel.: +(34)-91-3941840

Abstract: Treatment of a N-2-pyridyl-β-ketoamide precursor with bromine afforded the first example of the 3-aryl(α-hydroxy)methylenelimidazo[1,2-*a*]pyridin-2(3*H*)-one framework. This transformation proceeded through a domino process comprising an initial bromination, cyclization via an intramolecular S_N reaction, and a final keto-enol tautomerism, and allows generation of the fused heterocyclic system and installation of the acyl substituent in a single operation.

Keywords: nitrogen heterocycles; domino reactions; bromination; intramolecular S_N displacement

Citation: Tenti, G.; Cores, Á.; Ramos, M.T.; Menéndez, J.C. (*E*)-3-((2-Fluorophenyl)(hydroxy)methylene)-imidazo[1,2-*a*]pyridin-2(3*H*)-one. *Molbank* **2021**, *2021*, M1212. https://doi.org/10.3390/M1212

Academic Editors: Dimitrios Matiadis and Eleftherios Halevas

Received: 28 April 2021
Accepted: 8 May 2021
Published: 14 May 2021

Publisher's Note: MDPI stays neutral with regard to jurisdictional claims in published maps and institutional affiliations.

1. Introduction

Fused heterocyclic compounds bearing bridgehead nitrogen atoms are very relevant in the construction of diversity-oriented libraries. Imidazo[1,2-*a*]pyridine is one of these frameworks, and has shown great promise in the treatment of cancer due to the ability of many of its derivatives to inhibit a variety of kinases [1,2]. Moreover, several drugs currently in the market contain an imidazo[1,2-*a*]pyridine core, including the hypnotic zolpidem [3], the anxyolitic alpidem [4], the gastroprotective drug zolimidine [5], and the cardiotonic olprinone [6] (Figure 1). The importance of this ring system has led to much interest in its synthesis [7] and functionalization [8]. Imidazo[1,2-*a*]pyridin-2(3*H*)-one, one of its derivatives, is less well-known but has nevertheless shown some relevant pharmacological activities [9].

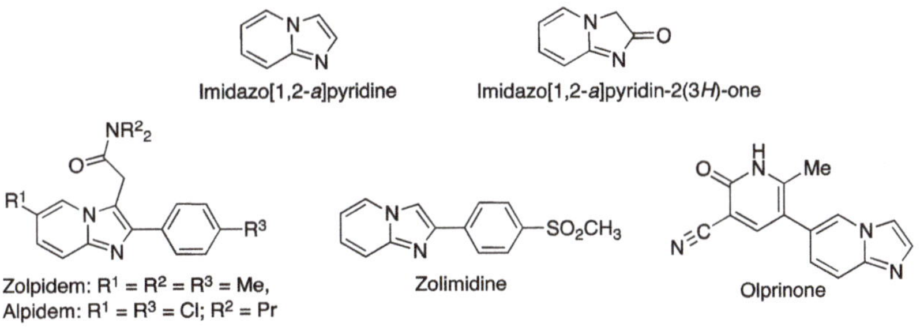

Figure 1. Structures of imidazo[1,2-*a*]pyridine, imidazo[1,2-*a*]pyridin-2(3*H*)-one and some relevant imidazo[1,2-*a*]pyridines.

In spite of the potential importance of the imidazo[1,2-*a*]pyridin-2(3*H*)-one framework, some of its derivatives have not been accessible to date. In particular, 3-acyl derivatives have been prepared only in one occasion, by acylation of an imidazo[1,2-*a*]pyridin-1-ium substrate (Scheme 1) [10]. This method yields compounds in mesoionic form, which was suitable for the purpose of the researchers, who aimed at the discovery of mesoionic nicotinic acetylcholine antagonists acting as insecticides [11], but it lacks generality.

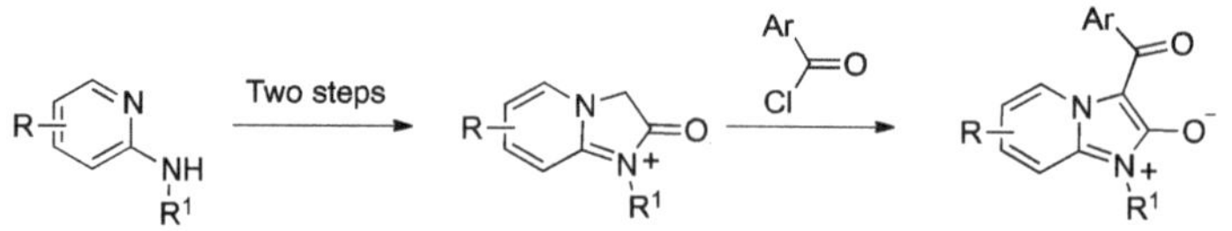

Scheme 1. Precedent for the synthesis of mesoionic 3-acyl-imidazo[3,2-*a*]pyridinium derivatives.

In this context, we describe here the synthesis of a representative of the 3-acylimidazo-[1,2-*a*]pyridin-2(3*H*)-one class of compounds from very simple starting materials and catalysts, using as the key step a domino reaction that generates the fused heterocyclic system and installs the acyl substituent in a single operation.

2. Results and Discussion

The synthesis of our target compound **5** is summarized in Scheme 2. The starting material **1**, a derivative of ethyl 3-oxo-3-phenylpropanoate, was treated with 2-aminopyridine in refluxing toluene containing acetic acid as catalyst to yield the β-ketoamide **3**. Its halogenation with bromine in dichloromethane afforded **5** via a domino sequence of reactions that comprised the initial halogenation of **3** at its central carbon, followed by cyclization by intramolecular S_N2 displacement of bromide anion and a final keto-enol tautomerism.

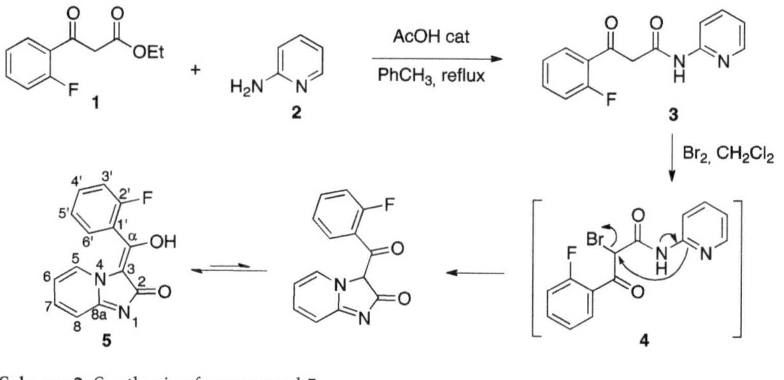

Scheme 2. Synthesis of compound **5**.

The enol structure of the imidazo[1,2-*a*]pyridin-2(3*H*)-one derivative **5** was established by the absence of signals ascribable to a proton neighboring two carbonyls or its corresponding carbon and also to a ketone carbonyl, all of which would be expected from a dicarbonyl structure. On the other hand, an enol proton was observed at 12.28 ppm, and the presence of quaternary carbons at 104.7 (C-3) and 175.3 ppm (C-α), which showed a correlation with H-6′ in the HMBC experiment, also support the enol tautomeric form. The alternative lactim tautomer was discarded because of the absence of a ketone carbonyl at ca. 190 ppm. The presence of a doublet at ca. 9.9 ppm with coupling constant 7 Hz, assigned to H-5, can be ascribed to the influence of the diamagnetic anisotropy cone of the arylmethylene substituent on the H-5 proton, and confirms the (*E*) configuration of the double bond. The assignment of the ^{1}H- and ^{13}C-NMR data was aided by the ^{19}F-^{13}C couplings, which allowed the unequivocal identification of the fluorophenyl ring carbons and also of the corresponding protons via H-C correlation 2D-NMR experiments (Supplementary Materials). Because H-5 had been assigned as described above, 2D-NMR data also enabled identification of the signals due to the H and C atoms at the 5, 6, 7, 8, and 8a positions. Copies of NMR spectra can be found in the Supporting Information.

3. Materials and Methods

General experimental information. All reagents (Sigma-Aldrich, Madrid, Spain; Fischer Chemical, Madrid, Spain; Alpha Aesar, Kändel, Germany) and solvents (Scharlau, Barcelona, Spain; Fischer Chemical, Madrid, Spain) were of commercial quality and were used as received. Reactions were monitored by thin layer chromatography on aluminum plates coated with silica gel and fluorescent indicator (Merck, Madrid, Spain) acherey-Nagel Xtra SIL G/UV254. Melting points were determined using a Stuart Scientific apparatus, SMP3 Model, and are uncorrected. Infrared spectra were recorded with an Agilent Cary630 FTIR spectrophotometer (Madrid, Spain) working by attenuated total reflection (ATR), with a diamond accessory for solid and liquid samples. NMR spectroscopic data were recorded using a Bruker Avance 250 spectrometer (Rivas-Vaciamadrid, Spain) operating at 250 MHz for ^{1}H-NMR and 63 MHz for ^{13}C-NMR maintained by the NMR facility of Universidad Complutense (CAI de Resonancia Magnética Nuclear, Madrid, Spain); chemical shifts are given in ppm and coupling constants in Hertz. ^{1}H- and ^{13}C-NMR assignments were supported by 2D-NMR experiments and are in agreement with simulations performed with MestreNova and ChemDraw Pro. Elemental analyses were determined by the microanalysis facility of Universidad Complutense (CAI de Microanálisis Elemental, Madrid, Spain), using a Leco 932 combustion microanalyzer.

3-(2-Fluorophenyl)-3-oxo-*N*-(pyridin-2-yl)propanamide (3). A solution of ethyl 3-(2-fluorophenyl)-3-oxopropanoate (631 mg, 3.0 mmol), 2-aminopyridine (310 mg, 3.3 mmol), and glacial acetic acid (0.3 mL) in toluene (7.5 mL) was heated under reflux with a Dean-Stark trap for 12 h. After completion of the reaction (checked by TLC), the solvent was evaporated under reduced pressure and the crude residue was purified by crystallization from ethyl ether giving **3** as a yellow solid (745 mg, 96%) comprised of a mixture of the β-ketoamide and its tautomeric enol form in a ratio of 1.4:1. MP: 142 °C. ^{1}H-NMR (250 MHz, DMSO-d_6) δ 14.39 (s, 1H, enol), 10.88 (s, 1H, enol), 10.71 (s, 1H, β-ketoamide), 8.38–8.28 (m, 1H, enol + β-ketoamide), 8.14–8.02 (m, 1H, enol + β-ketoamide), 7.93–7.50 (m, 3H, enol + β-ketoamide), 7.44–7.29 (m, 2H, enol + β-ketoamide), 7.12 (m, 1H, enol + β-ketoamide), 6.27 (s, 1H, enol), 4.16 (s, 2H, β-ketoamide) ppm. ^{13}C-NMR (63 MHz, DMSO-d_6) δ 192.13 (C_q), 192.08 (C_q), 171.29 (C_q), 166.18 (C_q), 164.93 (C_q), 163.30 (C_q), 161.91 (C_q), 159.25 (C_q), 157.89 (C_q), 151.87 (C_q), 151.52 (C_q), 148.21 (CH), 148.05 (CH), 138.30 (CH), 135.66 (CH), 135.51 (CH), 132.87 (CH), 132.73 (CH), 130.48 (CH), 130.45 (CH), 128.83 (CH), 128.80 (CH), 124.95 (CH), 124.90 (CH), 124.85 (CH), 124.69 (C_q), 121.79 (C_q), 121.63 (C_q), 119.74 (CH), 119.56 (CH), 117.15 (CH), 116.83 (CH), 116.79 (CH), 116.47 (CH), 114.24 (CH), 113.37 (CH), 95.05 (CH), 94.84 (CH), 51.78 (CH_2), 51.68 (CH_2). ESI-MS: $[M]^+$ 258. IR (neat, cm^{-1}): 3184, 2986, 1629, 16191579, 1536, 1438, 1400, 1307, 1297, 1195, 775. Elemental analysis calcd (%) for $C_{14}H_{11}FN_2O_2$: C, 65.11; H, 4.29; N, 10.85; found: C, 65.41; H, 4.52; N, 11.02.

(*E*)-3-((2-Fluorophenyl)(hydroxy)methylene)imidazo[1,2-*a*]pyridin-2(3*H*)-one (5). To a solution of compound **3** (258 mg, 1.0 mmol) in CH_2Cl_2 (5 mL) was added dropwise a solution of Br_2 (160 mg, 1.0 mmol) in CH_2Cl_2 (3 mL) at 0 °C. The reaction mixture was warmed to room temperature and stirred further for an additional 90 min. Then the mixture was washed with a saturated solution of $NaHCO_3$ and the organic layer was evaporated under reduced pressure giving a solid crude that was recrystallized from Et_2O to give compound **5** as a pale yellow solid (231 mg; 91%). MP: 167–168 °C. ^{1}H-NMR (250 MHz, DMSO-d_6) δ 12.28 (s, 1H, OH), 9.92 (d, *J* = 7.0 Hz, 1H, H-5), 7.79 (t, *J* = 8.0 Hz, 1H, H-7), 7.49–7.28 (m, 4H, H-6,8,4′,6′), 7.25–7.10 (m, 2H, H-3´,5) ppm. ^{13}C NMR (63 MHz, DMSO-d_6) δ 175.35 (C_q, C-α), 159.10 (d, *J* = 246.0 Hz, C_q, C-2′), 158.92 (C_q, C-3), 137.75 (C_q, C-9), 134.11 (CH, C-7), 130.90 (d, *J* = 8.2 Hz, CH, C-4′), 129.70 (d, *J* = 4.0 Hz, CH, C-6′), 129.26 (d, *J* = 17.1 Hz, C_q, C-1′), 129.05 (CH, C-7), 124.01 (d, *J* = 3.2 Hz, CH, C-5′), 116.41 (CH, C-6), 115.33 (d, *J* = 21.9 Hz, CH, C-3′), 108.21 (CH, C-8), 104.73 (C_q, C-3) ppm. ESI-MS: $[M]^+$ 256. IR (neat, cm^{-1}): 3015, 2969, 2841, 1699, 1686, 1610, 1569, 1438, 1307, 1257, 1212, 1139, 752. Elemental analysis calcd (%) for: $C_{14}H_9FN_2O_2$: C, 65.62; H, 3.54; N, 10.93; found: C, 65.81; H, 3.82; N, 11.20.

4. Conclusions

A N-2-pyridyl-β-ketoamide was shown to be a suitable precursor to the hitherto unknown 3-arylmethyleneimidazo[1,2-*a*]pyridin-2(3*H*)-one framework via a domino process comprising bromination, intramolecular nucleophilic substitution, and keto-enol tautomerism stages.

Supplementary Materials: The following are available online: copies of spectra of compounds **3** and **5**, including 2D-NMR correlation experiments.

Author Contributions: Conceptualization, G.T., Á.C., M.T.R., and J.C.M.; Methodology, G.T., and Á.C.; Writing—original draft preparation, J.C.M.; Writing—review and editing, G.T., Á.C., M.T.R., and J.C.M.; Supervision, M.T.R. and J.C.M.; Funding acquisition, J.C.M. All authors have read and agreed to the published version of the manuscript.

Funding: This research was funded by Ministerio de Ciencia, Innovación y Universidades (grant RTI2018-097662-B-I00), Comunidad de Madrid (postdoctoral contract to Á.C.) and Universidad Complutense (predoctoral contract to G.T.). The APC was not funded.

Institutional Review Board Statement: Not applicable.

Informed Consent Statement: Not applicable.

Data Availability Statement: Data are available from the authors, upon request.

Conflicts of Interest: The authors declare no conflict of interest.

References

1. Goel, R.; Luxami, V.; Paul, K. Imidazo[1,2-*a*]pyridines: Promising drug candidates for antitumor therapy. *Curr. Top. Med. Chem.* **2016**, *16*, 3590–3616. [CrossRef] [PubMed]
2. Yu, Y.; Han, Y.; Zhang, F.; Gao, Z.; Zhu, T.; Dong, S.; Ma, M. Design, synthesis, and biological evaluation of imidazo[1,2-*a*]pyridine derivatives as novel PI3K/mTOR dual inhibitors. *J. Med. Chem.* **2020**, *63*, 3028–3046. [CrossRef] [PubMed]
3. Foda, N.H.; Ali, S.M. Zolpidem tartrate. *Profiles Drug Subst. Excip. Relat. Methodol.* **2012**, *37*, 413–438. [PubMed]
4. Skolnick, P. Anxioselective anxiolytics: On a quest for the holy grail. *Trends Pharmacol. Sci.* **2012**, *33*, 611–620. [CrossRef] [PubMed]
5. Das, D.; Bhutia, Z.; Panjikar, P.C.; Chatterjee, A.; Banerjee, M. A simple and efficient route to 2-arylimidazo[1,2-*a*]pyridines and zolimidine using automated grindstone chemistry. *J. Heterocycl. Chem.* **2020**, *57*, 4099–4107. [CrossRef]
6. Mizushige, K.; Ueda, T.; Yukiiri, K.; Suzuki, H. Olprinone: A phosphodiesterase III inhibitor with positive inotropic and vasodilator effects. *Cardiovasc. Drug Rev.* **2006**, *20*, 163–174. [CrossRef] [PubMed]
7. Kumar Bagdi, A.; Santra, S.; Monir, K.; Hajra, A. Synthesis of imidazo[1,2-*a*]pyridines: A decade update. *Chem. Commun.* **2015**, *51*, 1555–1575. [CrossRef] [PubMed]
8. Tashrifi, Z.; Mohammadi-Khanaposhtani, M.; Larijani, B.; Mahdav, M. C3-functionalization of imidazo[1,2-*a*]pyridines. *Eur. J. Org. Chem.* **2020**, 269–284. [CrossRef]
9. Hemasrilatha, S.; Sruthi, K.; Manjula, A.; Harinadha Babu, V.; Vittal Rao, B. Synthesis and anti-inflammatory activity of imidazo[1,2-*a*]pyridinyl/pyrazinyl benzamides and acetamides. *Indian J. Chem.* **2012**, *51B*, 981–987.
10. Heil, M.; Hoffmeister, L.; Webber, M.; Ilg, K.; Goergens, U.; Turberg, A. Preparation of Mesoionic Imidazopyridines for Use as Insecticides. WO Patent 2018192872, 25 October 2018.
11. Liu, Z.; Li, Q.X.; Song, B. Recent research progress in and perspectives of mesoionic insecticides: Nicotinic acetylcholine receptor inhibitors. *J. Agric. Food Chem.* **2020**, *68*, 11039–11053. [CrossRef] [PubMed]

Short Note

1-Methyl-8-phenyl-1,3-diazaspiro[4.5]decane-2,4-dione

Vasiliki Pardali, Sotirios Katsamakas, Erofili Giannakopoulou and Grigoris Zoidis *

Department of Pharmacy, Division of Pharmaceutical Chemistry, School of Health Sciences, National and Kapodistrian University of Athens, Panepistimiopolis Zografou, 15771 Athens, Greece; vasilikipard@pharm.uoa.gr (V.P.); sotikats@eex.gr (S.K.); evgian@pharm.uoa.gr (E.G.)
* Correspondence: zoidis@pharm.uoa.gr; Tel.: +30-210-7274809

Abstract: A simple, fast and cost-effective three-step synthesis of 1-methyl-8-phenyl-1,3-diazaspiro[4.5] decane-2,4-dione has been developed. The reactions described herein proceed readily, with high yields and no further purification. Therefore, the proposed method, with an overall yield of 60%, offers a facile pathway to the synthesis of N-1 monosubstituted spiro carbocyclic imidazolidine-2,4-diones (hydantoins), which constitute a privileged class of heterocyclic scaffolds with pharmacological interest.

Keywords: carbocyclic hydantoins; N-1 substituted hydantoin; spiro hydantoins; imidazolidine-2,4-diones; DFT calculations; stereochemistry; NMR; HRMS; GIAO; ring closing

Citation: Pardali, V.; Katsamakas, S.; Giannakopoulou, E.; Zoidis, G. 1-Methyl-8-phenyl-1,3-diazaspiro[4.5]decane-2,4-dione. *Molbank* **2021**, *2021*, M1228. https://doi.org/10.3390/M1228

Academic Editors: Dimitrios Matiadis and Eleftherios Halevas

Received: 18 May 2021
Accepted: 1 June 2021
Published: 4 June 2021

Publisher's Note: MDPI stays neutral with regard to jurisdictional claims in published maps and institutional affiliations.

1. Introduction

Hydantoins, or imidazolidine-2,4-diones, is a class of compounds first isolated as a product from allantoin hydrogenolysis in 1861 from the Nobel laureate Adolph von Baeyer. Since then, hydantoin derivatives have become increasingly important, with various applications across chemical and pharmaceutical industries. Although, the hydantoin ring itself possesses no biological activity, 5- and especially 5,5-substituted derivatives have a documented, wide range of therapeutic applications [1]. The most noticeable drugs in this class of compounds showcase anticonvulsant [2], antidiabetic [3], anticancer [4], antiarrhythmic [5] and anti-inflammatory [6] activity. For example, Figure 1 depicts three well-known hydantoins with medicinal applications. Phenytoin exhibits a regulatory effect on the central nervous system (CNS) and has been applied successfully to ameliorating epilepsy symptoms for more than 70 years now and as a treatment of neuropathic pain [7]. More recently, Dantrolene (Figure 1), has been used in the clinical treatment of malignant hyperthermia through the inhibition of abnormal Ca^{2+} release at the sarcoplasmic reticulum and physiological Ca^{2+} release from skeletal muscles [8], whilst another compound known as BIRT377 (Figure 1) demonstrates potent anti-inflammatory activity as an antagonist of lymphocyte function-associated antigen-1 (LFA-1) [9]. Therefore, hydantoins are catching the attention of both the medicinal and organic chemistry spectrums based on their facilitated, privileged pharmacological profile [10].

Our medicinal chemistry lab has an active research interest in the development of such molecules incorporating bulky lipophilic carbocyclic rings into the spiro hydantoin core structure. These specific analogs are highly functionalized 'building blocks', with significant antiviral and trypanocidal activity, suitable for further synthetic transformations [11].

The introduction of substituents at the N-3 position of the hydantoin ring may be accomplished easily using alkyl halides in an alkaline solution. However, the synthesis of N-1 monosubstituted hydantoins cannot be achieved through direct alkylation unless the N-3 nitrogen is protected. The specified reactivity is explained due to the more activated N-3 position with the two neighboring activating carbonyl groups. In this paper, we report a simple, fast and effective 3-steps synthesis of 1-methyl-8-phenyl-1,3-diazaspiro[4.5]decane-2,4-dione (**4**). This synthetic route affords pure products in very good yields (overall yield of 60%) that can be used without further purifications. Therefore, the proposed facile and cost-effective method can be generally applied for the synthesis of N-1 monosubstituted spiro carbocyclic hydantoins that are building blocks of high interest to medicinal chemists.

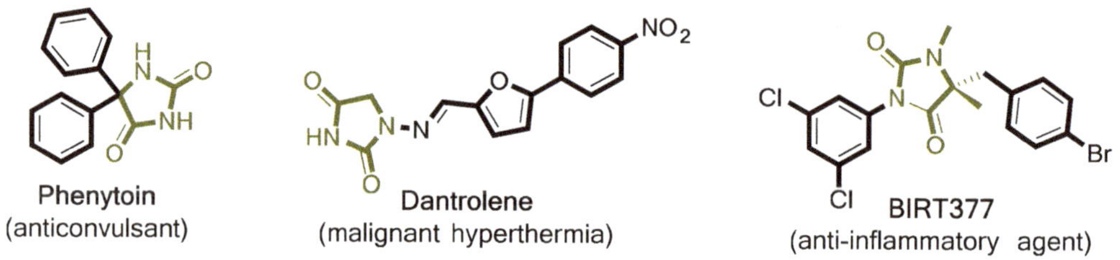
Phenytoin
(anticonvulsant)
Dantrolene
(malignant hyperthermia)
BIRT377
(anti-inflammatory agent)

Figure 1. Chemical structures of hydantoin-containing compounds with important biological activity.

2. Results

The target compound (**4**), is obtained by following the synthetic procedure shown in Scheme 1. The key intermediate α-amino nitrile hydrochloride (**2**) was obtained by a Strecker reaction of the bulky commercially available 4-phenylcyclohexan-1-one with NaCN and methylamine hydrochloride in a mixture of DMSO/H_2O. After workup, the resulting dry Et_2O solution of the free amino nitrile was treated with saturated ethanolic hydrochloric acid solution under ice cooling. 1-(Methylamino)-4-phenylcyclohexane-1-carbonitrile hydrochloride (**2**) was then treated with potassium cyanate (KOCN) in the presence of acetic acid and water to yield the corresponding ureido derivative (**3**). Cyclization of the latter with sodium hydride (60% NaH in mineral oil) in dry DMF and subsequent acid hydrolysis led to the target 1-methyl-8-phenyl-1,3-diazaspiro[4.5]decane-2,4-dione (**4**).

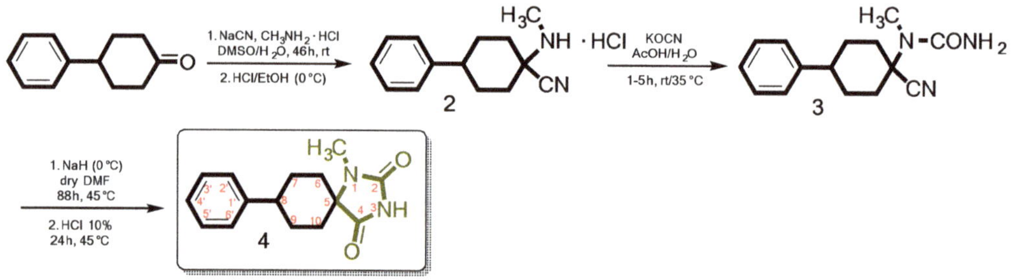
1. NaCN, CH3NH2 · HCl
DMSO/H2O, 46h, rt
2. HCl/EtOH (0 °C)
KOCN
AcOH/H2O
1-5h, rt/35 °C
1. NaH (0 °C)
dry DMF
88h, 45 °C
2. HCl 10%
24h, 45 °C

Scheme 1. The proposed facile and cost-effective method applied for the synthesis of N-1 monosubstituted spiro carbocyclic hydantoins.

For the target compound (**4**), a structure optimization step took place in order to determine its energy minima conformation. Starting from the selection of the cyclohexane, two more stable conformations (i.e., chair and twisted boat) [12], we drew the possible four more favorable structures that underwent the ab initio calculation (Figure 2). Although the twisted boat conformation of cyclohexane bears a 23.01 kJ/mol higher energy, it was nonetheless included in our experiments. Additionally, in all of the structures, the phenyl ring always occupies always the equatorial position since the axial position would introduce additional hindering effects on the respective conformations, resulting in even higher energies. Based on our calculations (Table 1), the energy minima for compound (**4**) resulted in being the conformation **B** (0 kJ/mol), while on ascending order the rest follow with the second being conformation **A** (9.43 kJ/mol), third being conformation **C** (26.08 kJ/mol) and fourth being conformation **D** (30.83 kJ/mol). Additional information upon all minimized conformers regarding bond lengths, angles and dihedrals can be found on Supplementary Tables S1–S4.

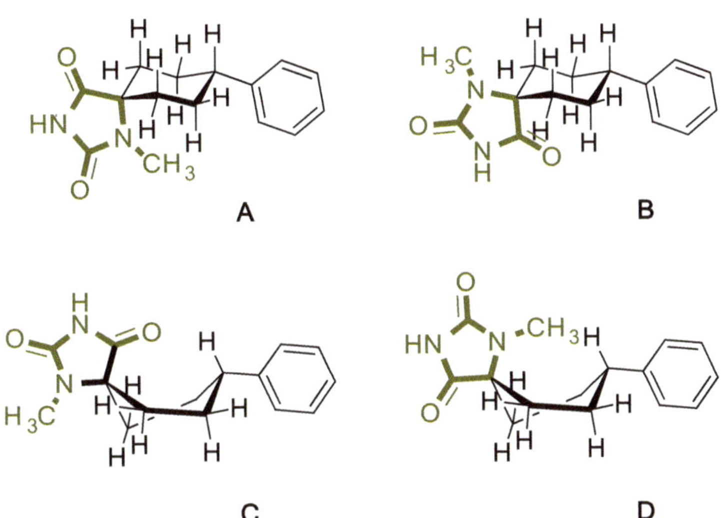

Figure 2. Chemical structure representation of compound (**4**) conformations (**A–D**).

Table 1. DFT summary of the calculated **4** conformations **A–D**.

Compound	Conformation A	Conformation B	Conformation C	Conformation D
4				
Charge	0	0	0	0
Spin	Singlet	Singlet	Singlet	Singlet
Solvation	None	None	None	None
E(RB3LYP)	-842.498395 Hartree	-842.501986 Hartree	-842.492052 Hartree	-842.490242 Hartree
RMS Gradient Norm	-	-	-	-
Imaginary Freq	-	-	-	-
Dipole Moment	3.364844	3.032529 Debye	3.042446 Debye	3.364416 Debye
Point Group	C1	C1	C1	C1

Based on the NMR of compound (**4**), provided on the supporting information, stereochemistry as determined from the data for the 3D structure with HMBC and NOESY (Figure 3a), correlations verify that the experimentally obtained structure matches the calculated energy minima conformation **B**. On the contrary Gauge-Independent Atomic Orbital (GIAO), theoretical calculations predicted similar signals for conformations **A–D** (Figure 3b) when referred to the hydantoin and the phenyl rings, respectively. For the cyclohexane ring conformations, **A** and **B** represent a more relatively true estimation, with two signals for carbons $C_{7,9}$ and $C_{6,10}$.

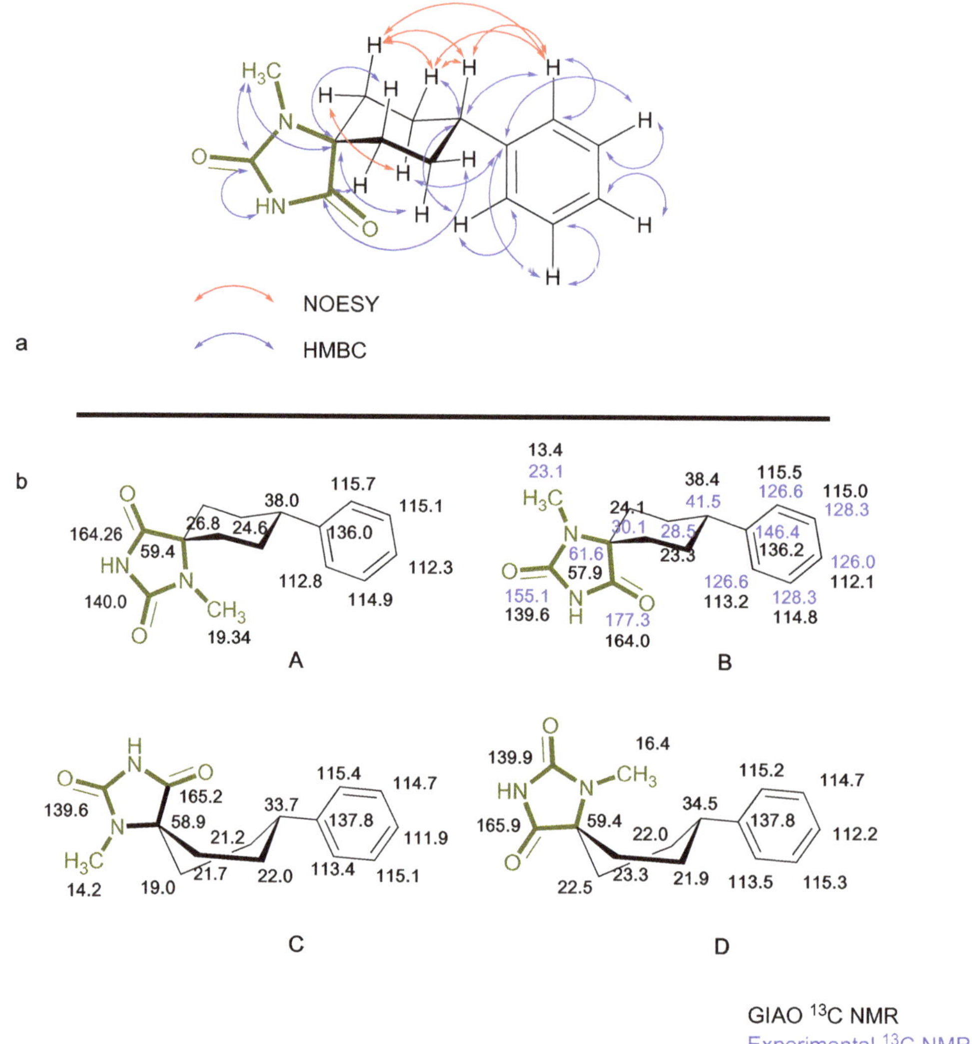

Figure 3. (**a**) Representation of experimental NMR data for (**4**) with HMBC correlations illustrated as blue two-way bent arrows and NOESY illustrated as red two-way bent arrows. (**b**) GIAO predicted ^{13}C NMR chemical shifts of all conformers with black writing and experimental shifts for the assigned conformer **B** in a blue color.

3. Materials and Methods

3.1. Chemistry

Materials, apparatuses and techniques for the experimental part are as follows. Melting points were determined using a Büchi capillary apparatus and are uncorrected. NMR experiments were performed to elucidate the structure and determine the purity of the newly synthesized compounds. ^{1}H NMR and 2D NMR spectra (COSY, HSQC-DEPT, HMBC) were recorded on a Bruker Ultrashield™ Plus Avance III 600 spectrometer (150.9 MHz, ^{13}C NMR). Chemical shifts δ *(delta)* are reported in parts per million (ppm) downfield from the NMR solvent, with the tetramethylsilane or solvent (DMSO-d_6) as internal standard. Data processing including Fourier transformation, baseline correction, phasing, peak peaking and integrations were performed using MestReNova software v.12.0.0. Splitting patterns are designated as follows: s, singlet; d, doublet; t, triplet; dd, doublet of doublets; td, triplet of doublets; m, multiplet; complex m, complex multiplet. Coupling constants (J) are expressed in units of Hertz (Hz). The spectra were recorded at 293 K (20 °C) unless otherwise specified. The solvent used to obtain the spectra was deuterated DMSO, DMSO-d_6 (quin, 2.50 ppm, ^{1}H NMR; septet, 39.52 ppm, ^{13}C NMR). IR spectra were recorded on a Perkin Elmer Spectrum BX FT-IR FTIR Spectrophotometer. Analytical thin-layer chromatography (TLC) was used to monitor the progress of the reactions, as well as to authenticate the compounds. TLCs were conducted on and precoated with normal-phase silica gel, aluminum sheets (Silica gel 60 F$_{254}$, Merck) (layer thickness 0.2 mm), precoated with reverse-phase silica gel, aluminum sheets (Silica gel 60 RP-18 F$_{254}$s, Merck) and precoated aluminum oxide plates (TLC Aluminium oxide 60 F$_{254}$, neutral). Developed plates were examined under a UV light source, at wavelengths of 254 nm or after being stained by iodine vapors. The Retention factor (R_f) of the newly synthesized compounds, which equals to the distance migrated over the total distance covered by the solvent, was also measured on the chromatoplates. Elemental analyses (C, H, N) were performed by the Service Central de Microanalyse at CNRS (Paris, France), and were within ±0.4% of the theoretical values. Elemental analysis results for the tested compounds correspond to >95% purity. The HRMS spectra were acquired in the negative ionization mode, employing a QTOF-MS (Maxis Impact, Bruker Daltonics, Bremen, Germany) using a resolving power of 40,000. The commercial reagents were purchased from Alfa Aesar, Sigma-Aldrich and Merck, and were used without further purification. Solvent abbreviations: ACN, acetonitrile; AcOEt, ethyl acetate; Et$_2$O, diethyl ether; EtOH, ethanol; MeOH, methanol.

3.2. Computational

Density Functional Theory (DFT) calculations are as follows. All in silico calculations were carried out on a typical desktop PC running a Windows 10 64-bit operating system (Intel i7 3.4 GHz CPU processors, RAM 32 GB), in gas phase unless otherwise specified, using the G09W [13] software package. The hybrid DFT method with Becke's [14] three-parameter functional and the nonlocal correlation provided by the Lee, Yang and Parr expression (B3LYP) [15] was used for optimization, employing the 6−31+G(d,p) basis set [16,17]. Single-point calculations of all structures were also carried out using the same basis set [16,17], following the optimized-geometries step. NMR shielding tensors computed with the GIAO method [18–22] on default parameters for degeneracy tolerance (0.05) in DMSO solvent, including spin–spin coupling constants [23–27].

3.3. Synthesis

1-(Methylamino)-4-phenylcyclohexane-1-carbonitrile hydrochloride (**2**): To a stirred suspension of sodium cyanide (843 mg, 17.2 mmol) and methylamine hydrochloride (1.16 g, 17.2 mmol) in 12 mL of DMSO/H$_2$O 9:1 (*v/v*), a solution of 4-phenylcyclohexanone (3.0 g, 17.2 mmol) in DMSO (24 mL) was added in one portion. The reaction mixture was stirred for 46 h at rt, poured into 130 mL of ice–water mixture and extracted with AcOEt (3 × 60 mL). The combined organic phases were washed with brine (2 × 60 mL), dried with anh. Na$_2$SO$_4$ and the solvent was evaporated under reduced pressure. The residue was dissolved in Et$_2$O (80 mL) and treated dropwise with an ethanolic solution saturated with gaseous hydrochloric acid to pH ~2 under ice bath. The precipitate formed was filtered off in vacuo, washed with small portions of dry Et$_2$O and dried over P$_2$O$_5$ to afford the title compound **2** as a white crystalline solid (3.55 g, 80%). This intermediate was used for the next reaction without further purification.

1-(1-Cyano-4-phenylcyclohexyl)-1-methylurea (**3**): To a stirred solution of the carbonitrile **2** (3.1 g, 12.4 mmol) in 20 mL acetic acid, a solution of potassium cyanate (2.01 g, 24.8 mmol) in 3 mL H$_2$O was added. After stirring for 1 h at 35 °C, the reaction mixture was poured into 70 mL H$_2$O and extracted with CHCl$_3$ (3 × 50 mL). The combined organic layers were washed with H$_2$O (3 × 50 mL) and brine (2 × 50 mL), dried with anh. Na$_2$SO$_4$ and the solvents were evaporated to dryness under reduced pressure to afford the title compound **3** as a white solid (2.97 g, 93%). This intermediate was used for the next reaction without further purification.

1-Methyl-8-phenyl-1,3-diazaspiro[4.5]decane-2,4-dione (**4**): A stirred solution of **3** (2.9 g, 11.3 mmol) in 40 mL dry DMF was cooled in an ice bath and sodium hydride (353 mg, 14.7 mmol, 60% dispersion in mineral oil) was added portion-wise. After 4 d of stirring at 45 °C under Argon, the mixture was treated with a solution of HCl 10% (96 mL) and stirring was continued for 24 h at 45 °C. After this time, the reaction mixture was poured into 400 mL of ice–water mixture and extracted with $CHCl_3$ (4 × 200 mL). The combined organic extracts were washed with H_2O (3 × 250 mL) and brine (2 × 250 mL), dried with anh. Na_2SO_4 and the solvent was evaporated in vacuo. The remaining solid was treated with Et_2O and *n*-pentane to give the desired compound 4 as a pale yellow crystalline product. (2.44 g, 79%); Rf = 0.34 (CH_2Cl_2/AcOEt 5:1); mp 211–214 °C (AcOEt/dry Et_2O-*n*-pentane).

^{1}H NMR (600 MHz, DMSO-d_6) δ 1.70-1.75 (m, 4H, H_6, H_7, H_9, H_{10}), 1.93 (td, 2H, J = 12.8, 4.3 Hz, H_6, H_{10}), 2.18 (qd, 2H, J = 13.1, 3.9 Hz, H_7, H_9), 2.57 (tt, 1H, J = 12.5, 3.5 Hz, H_8), 2.73 (s, 3H, CH_3), 7.19 (m, 1H, $H_{4'}$), 7.24 (m, 2H, $H_{2'}$, $H_{6'}$), 7.30 (t, 2H, J = 7.6 Hz, $H_{3'}$, $H_{5'}$), 10.73 (s, 1H, H_3) ppm.

^{13}C NMR (150 MHz, DMSO-d_6) δ 23.1 (CH_3), 28.5 (C_7, C_9), 30.1 (C_6, C_{10}), 41.5 (C_8), 61.6 (C_5), 126.0 ($C_{4'}$), 126.6 ($C_{2'}$, $C_{6'}$), 128.3 ($C_{3'}$, $C_{5'}$), 146.4 ($C_{1'}$), 155.1 (C_2=O), 177.3 (C_4=O) ppm.

IR (*mull*) ν 3143.9 (>N-H)$_{amide}$, 2923.6 (C-H)$_{sp}{}^2$, 2856.1 (C-H)$_{sp}{}^3$, 2723.2 (C-H)$_{sp}{}^3$, 1764.6 (C=O)$_{amide}$, 1707.5 (C=O)$_{amide}$, 1459.5 (C=C)$_{Ar}$, 1406.5 (C=C)$_{Ar}$, 1376.8 (C-H)$_{methyl}$, 1141.0 (C-N)$_{methyl}$, 1118.9 (C-N)$_{spiro}$, 1067.3 (>CH_2) cm^{-1}.

Anal. Calcd for $C_{15}H_{18}N_2O_2$: C, 69.74; H, 7.02; N, 10.84. Found: C, 69.90; H, 7.30; N, 10.50. HRMS *m/z* calc. for $C_{15}H_{18}N_2O_2$ [M − H]$^+$ 257.1285, obtained 257.1287.

4. Conclusions

1-Methyl-8-phenyl-1,3-diazaspiro[4.5]decane-2,4-dione (**4**) was obtained through a facile and cost-effective method that offers a general route for the synthesis of N-1 mono-substituted spiro carbocyclic hydantoins, which are heterocyclic scaffolds of great interest to the pharmaceutical industry. The reactions were executed under mild (rt to 45 °C) and sustainable (intermediates can be used without further purifications) conditions. The structure of compound (**4**) was determined using ^{1}H NMR, ^{13}C NMR, HSQC, HMBC, elemental analysis, FT-IR and HRMS, while stereochemistry was elucidated from the data for the 3D structure obtained by HMBC, NOESY and DFT energy minima/GIAO calculations.

Supplementary Materials: The following are available online, NMR spectra of (**4**) in DMSO-*d6* (600 MHz) (^{1}H, ^{13}C, COSY, HSQC-DEPT, HMBC, NOESY) p. S3, NMR spectra of (**4**) in CDCl$_3$ (400 MHz) (^{1}H, ^{13}C, COSY, HSQC-DEPT, HMBC) p.S9, HRMS Spectra of (**4**), p. S14, FT-IR Spectra of (**4**), p. S16, Table S1: Conformation **A** properties p. S17, Table S2: Conformation **B** properties p. S20, Table S3: Conformation **C** properties p. S23, Table S4: Conformation **D** properties p. S26.

Author Contributions: V.P.: investigation; S.K.: DFT calculations, data curation, writing—original draft preparation and editing; E.G.: editing; G.Z.: conceptualization, methodology, data curation, writing—original draft preparation and editing, supervision and funding acquisition. All authors have read and agreed to the published version of the manuscript.

Funding: G.Z. would like to thank 'Empirikion Foundation' for financial support. V.P. would like to thank the State Scholarships Foundation of Greece for providing her a Ph.D. fellowship (MIS-5000432, NSRF 2014–2020).

Data Availability Statement: Not applicable.

Acknowledgments: We thank Evangelos Gikas (Laboratory of Analytical Chemistry, Department of Chemistry, National and Kapodistrian University of Athens) for performing the HRMS spectra and Dimitra Benaki (Division of Pharmaceutical Chemistry, Department of Pharmacy, School of Health Sciences, National and Kapodistrian University of Athens) for performing the NMR experiments.

Conflicts of Interest: The authors declare no conflict of interest.

References

1. Cho, S.H.; Kim, S.-H.; Shin, D. Recent applications of hydantoin and thiohydantoin in medicinal chemistry. *Eur. J. Med. Chem.* **2019**, *164*, 517–545. [CrossRef] [PubMed]
2. Trišović, N.; Valentić, N.; Ušćumlić, G. Solvent effects on the structure-property relationship of anticonvulsant hydantoin derivatives: A solvatochromic analysis. *Chem. Cent. J.* **2011**, *5*, 62. [CrossRef]
3. Iqbal, Z.; Ali, S.; Iqbal, J.; Abbas, Q.; Qureshi, I.Z.; Hameed, S. Dual action spirobicycloimidazolidine-2,4-diones: Antidiabetic agents and inhibitors of aldose reductase-an enzyme involved in diabetic complications. *Bioorg. Med. Chem. Lett.* **2013**, *23*, 488–491. [CrossRef]
4. Zhang, M.; Liang, Y.R.; Li, H.; Liu, M.M.; Wang, Y. Design, synthesis, and biological evaluation of hydantoin bridged analogues of combretastatin A-4 as potential anticancer agents. *Bioorg. Med. Chem.* **2017**, *25*, 6623–6634. [CrossRef]
5. Czopek, A.; Byrtus, H.; Zagórska, A.; Siwek, A.; Kazek, G.; Bednarski, M.; Sapa, J.; Pawłowski, M. Design, synthesis, anticonvulsant, and antiarrhythmic properties of novel N-Mannich base and amide derivatives of β-tetralinohydantoin. *Pharmacol. Rep.* **2016**, *68*, 886–893. [CrossRef]
6. Wang, Z.D.; Sheikh, S.O.; Zhang, Y. A Simple Synthesis of 2-Thiohydantoins. *Molecules* **2006**, *11*, 739–750. [CrossRef]
7. Kopsky, D.J.; Keppel Hesselink, J.M. Phenytoin Cream for the Treatment of Neuropathic Pain: Case Series. *Pharmaceuticals* **2018**, *11*, 53. [CrossRef]
8. Hosoya, T.; Aoyama, H.; Ikemoto, T.; Hiramatsu, T.; Kihara, Y.; Endo, M.; Suzuki, M. Dantrolene analogues revisited: General synthesis and specific functions capable of discriminating two kinds of Ca^{2+} release from sarcoplasmic reticulum of mouse skeletal muscle. *Bioorg. Med. Chem.* **2003**, *11*, 663–673. [CrossRef]
9. Last-Barney, K.; Davidson, W.; Cardozo, M.; Frye, L.L.; Grygon, C.A.; Hopkins, J.L.; Jeanfavre, D.D.; Pav, S.; Stevenson, J.M.; Tong, L.; et al. Binding Site Elucidation of Hydantoin-Based Antagonists of LFA-1 Using Multidisciplinary Technologies: Evidence for the Allosteric Inhibition of a Protein−Protein Interaction. *J. Am. Chem. Soc.* **2001**, *123*, 5643–5650. [CrossRef]
10. Giannakopoulou, E.; Pardali, V.; Skrettas, I.; Zoidis, G. Transesterification instead of N-Alkylation: An Intriguing Reaction. *ChemistrySelect* **2019**, *4*, 3195–3198. [CrossRef]
11. Giannakopoulou, E.; Pardali, V.; Frakolaki, E.; Siozos, V.; Myrianthopoulos, V.; Mikros, E.; Taylor, M.C.; Kelly, J.M.; Vassilaki, N.; Zoidis, G. Scaffold hybridization strategy towards potent hydroxamate-based inhibitors of Flaviviridae viruses and Trypanosoma species. *Med. Chem. Commun.* **2019**, *10*, 991–1006. [CrossRef]
12. Dragojlovic, V. Conformational analysis of cycloalkanes. *ChemTexts* **2015**, *1*, 1–30. [CrossRef]
13. Frisch, M.J.; Trucks, G.W.; Schlegel, H.B.; Scuseria, G.E.; Robb, M.A.; Cheeseman, J.R.; Scalmani, G.; Barone, V.; Mennucci, B.; Petersson, G.A.; et al. *Gaussian 09, Revision E.01*; Gaussian, Inc.: Wallingford, CT, USA, 2009.
14. Becke, A.D. Density-functional thermochemistry. III. The role of exact exchange. *J. Chem. Phys.* **1993**, *98*, 5648–5652. [CrossRef]
15. Lee, C.; Yang, W.; Parr, R.G. Development of the Colle-Salvetti correlation-energy formula into a functional of the electron density. *Phys. Rev. B Condens. Matter* **1988**, *37*, 785–789. [CrossRef]
16. Clark, T.; Chandrasekhar, J.; Spitznagel, G.W.; Schleyer, P.V.R. Efficient diffuse function-augmented basis sets for anion calculations. III. The 3-21+G basis set for first-row elements, Li–F. *J. Comput. Chem.* **1983**, *4*, 294–301. [CrossRef]
17. McLean, A.D.; Chandler, G.S. Contracted Gaussian basis sets for molecular calculations. I. Second row atoms, Z=11–18. *J. Chem. Phys.* **1980**, *72*, 5639–5648. [CrossRef]
18. Wolinski, K.; Hinton, J.F.; Pulay, P. Efficient implementation of the gauge-independent atomic orbital method for NMR chemical shift calculations. *J. Am. Chem. Soc.* **2002**, *112*, 8251–8260. [CrossRef]
19. Cheeseman, J.R.; Trucks, G.W.; Keith, T.A.; Frisch, M.J. A comparison of models for calculating nuclear magnetic resonance shielding tensors. *J. Chem. Phys.* **1996**, *104*, 5497–5509. [CrossRef]
20. Ditchfield, R. Self-consistent perturbation theory of diamagnetism. *Mol. Phys.* **1974**, *27*, 789–807. [CrossRef]
21. McWeeny, R. Perturbation Theory for the Fock-Dirac Density Matrix. *Phys. Rev.* **1962**, *126*, 1028–1034. [CrossRef]
22. London, F. Théorie quantique des courants interatomiques dans les combinaisons aromatiques. *J. Phys. Radium* **1937**, *8*, 397–409. [CrossRef]
23. Deng, W.; Cheeseman, J.R.; Frisch, M.J. Calculation of Nuclear Spin-Spin Coupling Constants of Molecules with First and Second Row Atoms in Study of Basis Set Dependence. *J. Chem. Theory Comput.* **2006**, *2*, 1028–1037. [CrossRef]
24. Peralta, J.E.; Scuseria, G.E.; Cheeseman, J.R.; Frisch, M.J. Basis set dependence of NMR spin–spin couplings in density functional theory calculations: First row and hydrogen atoms. *Chem. Phys. Lett.* **2003**, *375*, 452–458. [CrossRef]
25. Barone, V.; Peralta, J.E.; Contreras, R.H.; Snyder, J.P. DFT Calculation of NMR JFF Spin−Spin Coupling Constants in Fluorinated Pyridines. *J. Phys. Chem. A* **2002**, *106*, 5607–5612. [CrossRef]
26. Helgaker, T.; Watson, M.; Handy, N.C. Analytical calculation of nuclear magnetic resonance indirect spin–spin coupling constants at the generalized gradient approximation and hybrid levels of density-functional theory. *J. Chem. Phys.* **2000**, *113*, 9402–9409. [CrossRef]
27. Sychrovský, V.R.; Gräfenstein, J.; Cremer, D. Nuclear magnetic resonance spin–spin coupling constants from coupled perturbed density functional theory. *J. Chem. Phys.* **2000**, *113*, 3530–3547. [CrossRef]

Short Note

4-Amino-2-(*p*-tolyl)-7*H*-chromeno[5,6-*d*]oxazol-7-one

Evangelia-Eirini N. Vlachou [1], Thomas D. Balalas [1], Dimitra J. Hadjipavlou-Litina [2] and Konstantinos E. Litinas [1,*]

[1] Laboratory of Organic Chemistry, Department of Chemistry, Aristotle University of Thessaloniki, 54124 Thessaloniki, Greece; e.e.vlachou@gmail.com (E.-E.N.V.); thombal@hotmail.com (T.D.B.)

[2] Department of Pharmaceutical Chemistry, School of Pharmacy, Faculty of Health Sciences, Aristotle University of Thessaloniki, 54124 Thessaloniki, Greece; hadjipav@pharm.auth.gr

[*] Correspondence: klitinas@chem.auth.gr; Tel.: +30-2310-997864

Abstract: The new 4-amino-2-(*p*-tolyl)-7*H*-chromeno[5,6-*d*]oxazol-7-one was successfully prepared through the Au/TiO_2-catalyzed $NaBH_4$ activation and chemoselective reduction of the new 4-nitro-2-(*p*-tolyl)-7*H*-chromeno[5,6-*d*]oxazol-7-one. The latter was synthesized by the one-pot tandem reactions of 6-hydroxy-5,7-dinitrocoumarin with *p*-tolylmethanol under Au/TiO_2 catalysis. The dinitrocoumarin was obtained by the nitration of 6-hydroxycoumarin with cerium ammonium nitrate (CAN). The structure of the synthesized compounds was confirmed by FT-IR, HR-MS, ^{1}H-NMR and ^{13}C-NMR analysis. Preliminary biological tests show low anti-lipid peroxidation activity for the title compound.

Keywords: Au-nanoparticles; $NaBH_4$; amino-substituted fused oxazolocoumarin; fused oxazolocoumarins; chemoselective reduction; *o*-hydroxynitrocoumarins

Citation: Vlachou, E.-E.N.; Balalas, T.D.; Hadjipavlou-Litina, D.J.; Litinas, K.E. 4-Amino-2-(*p*-tolyl)-7*H*-chromeno[5,6-*d*]oxazol-7-one. *Molbank* **2021**, *2021*, M1237. https://doi.org/10.3390/M1237

Academic Editors: Dimitrios Matiadis and Eleftherios Halevas

Received: 30 May 2021
Accepted: 11 June 2021
Published: 15 June 2021

Publisher's Note: MDPI stays neutral with regard to jurisdictional claims in published maps and institutional affiliations.

1. Introduction

Coumarin derivatives are widely distributed in nature, presenting interesting biological properties such as anticoagulant, anti-inflammatory, antivirus, anticancer, antioxidant or antidiabetic [1–7]. Fused coumarins also exhibit biological activity. Especially, fused oxazolocoumarins have been tested for their antioxidant [8], antimicrobial [9], anti-inflammatory [10], photosensitizing [11] or photoreleasing of aminolevulinic acid [12] activities. There are several methodologies for the synthesis of fused oxazolocoumarins. The condensation of *o*-aminohydroxycoumarins with aldehydes [9,13–15], acids [14], anhydrides [13,15]; or of *o*-amidohydroxycoumarins with anhydrides [16], $POCl_3$ [17] or P_2O_5 [18] led to those products. Furthermore, substituted fused oxazolocoumarins were synthesized by the reduction of 4-hydroxy-3-nitrosocoumarin in acetic anhydride in the presence of Pd/C [19], or of 6-hydroxy-4-methyl-5-nitrocoumarin acetate in acetic acid with iron powder [20], or of 3-hydroxy-3-nitrocoumarins in liquid carboxylic acids in the presence of Pd/C or PPh_3 and P_2O_5 [8]. Recently, we prepared oxazolocoumarins by one-pot tandem reactions of *o*-hydroxynitrocoumarins with benzyl alcohol in toluene under catalytical conditions using gold nanoparticles supported on TiO_2, by $FeCl_3$ or by silver nanoparticles supported on TiO_2 [21].

Aminocoumarins are valuable building blocks for the synthesis of fused pyridocoumarins presenting significant biological activities such as antibacterial [22], antifungal [23], antimalarial [24], antioxidant [25] and wound-healing [26]. Pyridocoumarins are prepared from aminocoumarins through the one-pot Povarov reactions with aromatic aldehydes and cyclic enol ethers [27], the reactions with vinyl ketones [28], or under Vilsmeier conditions [29] or with phenylacetylene and benzaldehydes under catalysis by I_2 [30] or by other Lewis acids [25,31]. The cycloisomerization of propargylaminocoumarins, prepared from aminocoumarins, followed by oxidation, led also to pyridocoumarins under catalysis by $AgSbF_6$ [32] or $BF_3 \cdot Et_2O$ [33] or Au/nanoparticles [34].

The need for the synthesis of new compounds, to probe novel biological activity containing a heterocyclic ring fused to the pyridocoumarin moiety, led us to the synthesis of amino-substituted fused oxazolocoumarins. In continuation of our interest on fused oxazolocoumarin [8,22] and pyridocoumarin [25,33,34] derivatives, we would like to report here the synthesis of novel amine **7**, through a selective reduction procedure, and the biological evaluation of the products. The reactions studied and the synthesized products are depicted in Scheme 1.

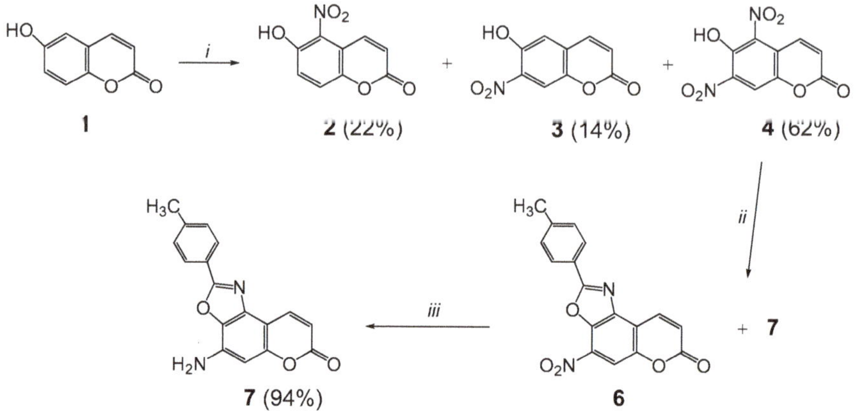

Scheme 1. Reagents and Conditions: (i) CAN (1 equiv.), CH_3CN, r.t. 30 min; (ii) p-tolylmethanol (5) (3 equiv.), Au/TiO_2 (4 mol%), toluene, sealed tube, 150 °C, 54 h; (iii) Au/TiO_2 (1 mol%), $NaBH_4$ (4 equiv.), MeOH, r.t. 1 h.

2. Results and Discussion

2.1. Synthesis

The starting material for this procedure was the 6-hydroxy-5,7-dinitrocoumarin (**4**), which was synthesized in 62% yield along with 6-hydroxy-5-nitrocoumarin (**2**) (22% yield) and 6-hydroxy-7-nitrocoumarin (**3**) (14% yield) by nitration of 6-hydroxycoumarin (**1**) with cerium ammonium nitrate (CAN) in CH_3CN at r.t., according to the literature [35]. In this paper, the authors obtained **3** in 50% yield using 1 equiv. of CAN, while by using 2 equiv. of CAN they isolated compound **3** in 74% yield along with compound **2** (12%). No evidence for the presence of the dinitro-derivative **4** was noticed. When we performed the above reaction with 0.5 equiv. of CAN, only compound **2** [36] (10 %) was isolated along with 85% of the starting compound **1**. The spectral data of compound **4** resemble well with that given in the literature [37], where the preparation was achieved by using nitric/acetic acids.

The reaction of **4** with *p*-tolylmethanol (**5**) in a sealed tube in toluene in the presence of Au/TiO_2 (4 mol%) at 150 °C led to 4-nitro-2-(*p*-tolyl)-7*H*-chromeno[5,6-*d*]oxazol-7-one (**6**) (45% yield) accompanied by 4-amino-2-(*p*-tolyl)-7*H*-chromeno[5,6-*d*]oxazol-7-one (**7**) (13%). This reaction was performed in analogy to our recent work on the synthesis of fused oxazolocoumarins by the treatment of *o*-hydroxynitrocoumarins with benzyl alcohol catalyzed by Au/TiO_2 or Ag/TiO_2 or $FeCl_3$ [21]. During this process, a simultaneous reduction of nitro- to amine-group and oxidation of benzyl alcohol to benzaldehyde occurred, followed by imine formation from the amine and benzaldehyde, cyclization by addition of hydroxy-group to imine and oxidation of the intermediate oxazoline to oxazole. The selective reduction of the 5-nitro group of coumarin in comparison to the 7-nitro group by the intermediate gold-hydride [21] could be attributed to a possible complexation of gold to the 3,4-double bond of coumarin. In the ^{1}H-NMR spectrum of **6**, there are two doublets at 6.42 (1 H, *J* = 9.6 Hz) and 8.28 (1 H, *J* = 9.6 Hz) for the 3-H and 4-H, respectively, and one singlet at 8.30 (1 H) for the 8-H. The chemical shift of 4-H (8.28 ppm) is downfield in comparison to 4-H (7.69 ppm) of compound **4** due possibly to de-shielding from the oxazole-ring. The p-tolyl-group gave rise to the two doublets at 7.35 (1 H, *J* = 7.9 Hz) and

8.15 (1 H, J = 7.9 Hz) and one singlet at 2.43 (3 H). The HR-MS is m/z [M + H]$^+$ calcd for $C_{17}H_{11}N_2O_5$: 323.2789, found: 323.2791.

The reduction of nitro-derivative **6** with NaBH$_4$ as hydride ion donor, in the presence of the catalyst Au/TiO$_2$, according to a recent publication for the use of Au-NPs in the reduction of nitroarenes to anilines [38], resulted to the chemoselective preparation of 4-amino-2-(*p*-tolyl)-7*H*-chromeno[5,6-*d*]oxazol-7-one (**7**) in 94% yield. This is a new compound with absorptions in FT-IR at 3446, 3356 cm^{-1} for the NH$_2$ group. There are two doublets at 6.29 (1 H, J = 9.6 Hz) and 8.26 (1 H, J = 9.6 Hz) for the 3-H and 4-H, respectively, in the ^{1}H-NMR spectrum of **7**, a broad singlet at 4.50 ppm for the NH$_2$ protons and one singlet at 6.61 (1 H) for the 8-H, see Supplementary Materials. This upfield shift is consistent with the structure of **7** with the oxazole-ring fused at the 5,6-position and the NH$_2$ group at the 7-position of the coumarin moiety. If the oxazole ring is at the 6,7-position and the amine group at the 5-position of the coumarin (in a structure isomeric to **7**), the 8-H would be expected to be above 7.0 ppm. In the case of 2-phenyl-6*H*-chromeno[6,7-*d*][1,3]oxazol-6-one the 8-H is at 7.54 ppm [21]. The p-tolyl group gives rise to two doublets at 7.36 (1 H, J = 7.9 Hz) and 8.15 (1 H, J = 7.9 Hz) and one singlet at 2.46 (3 H). In the ^{13}C-NMR, there is the upfield peak for the 8-C of the coumarin moiety at 98.1 ppm in comparison to the carbons of nitro-compound **6**, see Supplementary Materials. This peak is consistent with the analogous peak (98.9 ppm) for 7-aminocoumarin [39]. The HR-MS is m/z [M + Na]$^+$ calcd for $C_{17}H_{12}NaN_2O_3$: 315.2778, found: 315.2784.

2.2. Biology

Preliminary biological experiments were performed in vitro. Compounds **6** and **7** were tested as possible antioxidant agents and inhibitors of soybean lipoxygenase according to our previous published assays [10,25]. They did not present any interaction with DPPH at 100 μM after 20 and 60 min under the reported experimental conditions. The anti-lipid peroxidation activity was very low at 100 μM (less than 1% for compound **6** and 23% for compound **7**), as tested by the 2,2′-azobis(2-amidinopropane) dihydrochloride (AAPH) protocol. No inhibition of soybean lipoxygenase was observed.

3. Materials and Methods

3.1. Materials

All the chemicals were procured from either Sigma–Aldrich Co. or Merck & Co., Inc. (St. Louis, MO, USA) Melting points were determined with a Kofler hotstage apparatus and are uncorrected. IR spectra were obtained with a Perkin–Elmer Spectrum BX spectrophotometer as KBr pellets. NMR spectra were recorded with an Agilent 500/54 (DD2) (Santa Clara, CA, USA) (500 MHz and 125 MHz for ^{1}H and ^{13}C, respectively) using CDCl$_3$ as solvent and TMS as an internal standard. J values are reported in Hz. Mass spectra were determined with a LCMS-2010 EV Instrument (Shimadzu, Kyoto, Japan) under electrospray ionization (ESI) conditions. HRMS (ESI-MS) were recorded with a ThermoFisher Scientific model LTQ Orbitrap Discovery MS. Silica gel No. 60, Merck A.G. was used for column chromatography.

3.2. Synthesis of 6-Hydroxy-5,7-dinitrocoumarin (**4**)

Cerium ammonium nitrate (CAN) (1.69 g, 3.08 mmol) in acetonitrile (10 mL) was added in three portions over a period of 15 min to a solution of 6-hydroxycoumarin (**1**) (0.5 g, 3.08 mmol) in acetonitrile (10 mL) under stirring. The reaction mixture was then stirred for 30 min (TLC-monitored) and then quenched by pouring over ice (~50 g). It was then repeatedly extracted with ethyl acetate (3 × 10 mL). The combined extracts washed successively with sodium bisulfite solution, brine and water, and dried (Na$_2$SO$_4$). After evaporation, the residue was subjected to column chromatography [silica gel, hexane: ethyl acetate (1:1)] to give **2** and **3** as a mixture followed by the 6-hydroxy-5,7-dinitrocoumarin (**4**) (0.48 g, 62 % yield). The mixture of **2** and **3** were subjected to a second column chromatography [silica gel, dichloromethane] to give 6-hydroxy-5-nitrocoumarin (**2**) (0.14 g, 22 % yield) and 6-hydroxy-7-nitrocoumarin (**3**) (89 mg, 14% yield).

6-Hydroxy-5,7-Dinitrocoumarin (**4**): Red solid, m.p. 153–155 °C (dec) (EtOH), (lit. [37]: 155–157 °C).

6-Hydroxy-5-nitrocoumarin (**2**): Yellow solid, m.p. 159–161 °C (EtOH), (lit. [36]: 158–160 °C).

6-Hydroxy-7-nitrocoumarin (**3**): Yellow solid, m.p. 231–233 °C (EtOH), (lit. [36]: 232 °C).

3.3. Synthesis of 4-Nitro-2-(p-tolyl)-7H-chromeno[5,6-d]oxazol-7-one (6)

The 6 hydroxy 5,7 dinitrocoumarin (**4**) (100 mg, 0.40 mmol), *p*-tolylmethanol (**5**) (145.4 mg, 1.19 mmol), 1 % Au/TiO$_2$ [156.2 mg (1.56 mg Au, 0.00793 mmol, 2 mol%)] and toluene (4 mL) were added in a sealed tube. The resulted mixture was stirred at 150 °C for 54 h. After cooling, the catalyst was removed by filtration and the solvent was concentrated under reduced pressure. The residue was subjected to column chromatography [silica gel, hexane: ethyl acetate (2:1)] to give compound **6** (57 mg, 45 % yield) followed by the 4-amino-2-(*p*-tolyl)-7*H*-chromeno[5,6-*d*]oxazol-7-one (**7**) (15.2 mg, 13 % yield) and unreacted compound **4** (40 mg, 40 %).

4-Nitro-2-(p-tolyl)-7*H*-chromeno[5,6-*d*]oxazol-7-one (**6**): Light yellow solid, m.p. 90–92 °C (MeOH). IR (KBr): 3052, 2924, 2853, 1716 cm^{-1}. ^{1}H-NMR (500 MHz, CDCl$_3$) δ: 2.43 (s, 3H, CH$_3$), 6.42 (d, 1H, *J* = 9.6 Hz), 7.35 (d, 2H, *J* = 7.9 Hz), 8.15 (d, 2H, *J* = 7.9 Hz), 8.28 (d, 1H, *J* = 9.6 Hz), 8.30 (s, 1H). ^{13}C-NMR (125 MHz, CDCl$_3$) δ: 30.9, 111.1, 116.5, 117.5, 127.4, 127.67, 127.7, 129.9, 132.2, 136.8, 145.8, 146.0, 155.5, 160.6, 164.0. LC-MS (ESI): 320 [M − H]$^-$. HR-MS (ESI), (M.W.: 322): *m/z* [M + H]$^+$ calcd for C$_{17}$H$_{11}$N$_2$O$_5$: 323.2789, found: 323.2791.

3.4. Synthesis of 4-Amino-2-(p-tolyl)-7H-chromeno[5,6-d]oxazol-7-one (7)

The catalyst, 1% Au/TiO$_2$ [12.2 mg (0.12 mg Au, 0.0006 mmol, 1 mol%)], was placed in a 5 mL flask, followed by the addition of methanol (2 mL), nitro compound **6** (20 mg, 0.062 mmol) and NaBH$_4$ (gradual addition because of hydrogen release (9.4 mg, 0.25 mmol)). The reaction mixture was then stirred at room temperature for 1 h. After the completion of the reaction (TLC-monitored), the slurry was filtered under reduced pressure to remove the catalyst and washed with methanol (~5 mL). The filtrate was evaporated under vacuum to afford the corresponding 4-amino-2-(*p*-tolyl)-7*H*-chromeno[5,6-*d*]oxazol-7-one, (**7**) (17 mg, 94 % yield): Light yellow solid, m.p. 177–179 °C (hexane/ethyl acetate). IR (KBr): 3446, 3356, 2924, 2852, 1725, 1634 cm^{-1}. ^{1}H-NMR (500 MHz, CDCl$_3$) δ: 2.46 (s, 3H, CH$_3$), 4.50 (brs, 2H), 6.29 (d, 1H, *J* = 9.6 Hz), 6.61 (s, 1H), 7.36 (d, 2H, *J* = 7.9 Hz), 8.15 (d, 2H, *J* = 7.9 Hz), 8.26 (d, 1H, *J* = 9.6 Hz). ^{13}C-NMR (125 MHz, CDCl$_3$) δ: 31.0, 98.1, 111.4, 116.5, 117.4, 127.3, 127.7, 129.8, 129.9, 139.2 146.1, 146.7, 148.9, 156.1, 160.0, 164.7. LC-MS (ESI): 315 [M + Na]$^+$, 347 [M + Na + MeOH]$^+$. HR-MS (ESI), (M.W.: 292): *m/z* [M + Na]$^+$ calcd for C$_{17}$H$_{12}$NaN$_2$O$_3$: 315.2778, found: 315.2784.

3.5. Biological Experiments: In Vitro Assays

The compounds were dissolved in DMSO.

- Antilipid peroxidation: the AAPH protocol was followed [25].
- Lipoxygenase inhibition: according to our previous protocol [25].
- Antioxidant activity: interaction with the stable free radical DPPH (final concentration 0.05 mM) in ethanol absolute (final concentration of the tested compounds 0.1 mM) [25].

4. Conclusions

We demonstrated an efficient and chemoselective method for the synthesis of amino-substituted fused oxazolocoumarins using Au-NPs catalysis in the presence of NaBH$_4$ for the reduction of the corresponding nitro-substituted fused oxazolocoumarins. The preliminary biological assays pointed that compound **7** presents low anti-lipid peroxidation activity.

Supplementary Materials: The following are available online, NMR and LC-MS (ESI) spectra of compound 7.

Author Contributions: Conceptualization, writing—original draft preparation, supervision, K.E.L.; performed the biological tests, review and editing the manuscript, D.J.H.-L.; performed the experiments, E.-E.N.V.; performed experiments, editing, in part, the manuscript, T.D.B. All authors have read and agreed to the published version of the manuscript.

Funding: This research was funded by "Human Resources Development, Education and Lifelong Learning", EDBM103, "Synthesis of Fused Pyranoquinolinone Derivatives with possible Biological Interest" (MIS: 5066801) and "Support for researchers with emphasis on young researchers-cycle B", (NSRF 2014-2020), (KA1020216).

Data Availability Statement: The data presented in this study are available in this article.

Acknowledgments: "Human Resources Development, Education and Lifelong Learning", EDBM103, "Synthesis of Fused Pyranoquinolinone Derivatives with possible Biological Interest" (MIS: 5066801) and "Support for researchers with emphasis on young researchers-cycle B", (NSRF 2014-2020), (KA1020216).

Conflicts of Interest: The authors declare no conflict of interest.

References

1. Yu, D.; Suzuki, M.; Xie, L.; Morris-Natschke, S.L.; Lee, K.-H. Recent progress in the development of coumarin derivatives as potent anti-HIV agents. *Med. Res. Rev.* **2003**, *23*, 322–345. [CrossRef] [PubMed]
2. Fylaktakidou, K.; Hadjipavlou-Litina, D.; Litinas, K.; Nicolaides, D. Natural and Synthetic Coumarin Derivatives with Anti-Inflammatory/Antioxidant Activities. *Curr. Pharm. Des.* **2004**, *10*, 3813–3833. [CrossRef] [PubMed]
3. Lacy, A. Studies on Coumarins and Coumarin-Related Compounds to Determine their Therapeutic Role in the Treatment of Cancer. *Curr. Pharm. Des.* **2004**, *10*, 3797–3811. [CrossRef]
4. Medina, F.G.; Marrero, J.G.; Macías-Alonso, M.; González, M.C.; Córdova-Guerrero, I.; García, A.G.T.; Osegueda-Robles, S. Coumarin heterocyclic derivatives: Chemical synthesis and biological activity. *Nat. Prod. Rep.* **2015**, *32*, 1472–1507. [CrossRef] [PubMed]
5. Kubrak, T.; Podgórski, R.; Stompor, M. Natural and Synthetic Coumarins and their Pharmacological Activity. *Eur. J. Clin. Exp. Med.* **2017**, *15*, 169–175. [CrossRef]
6. Li, H.; Yao, Y.; Li, L. Coumarins as potential antidiabetic agents. *J. Pharm. Pharmacol.* **2017**, *69*, 1253–1264. [CrossRef]
7. Salehian, F.; Nadri, H.; Jalili-Baleh, L.; Youseftabar-Miri, L.; Bukhari, S.N.A.; Foroumadi, A.; Küçükkilinç, T.T.; Sharifzadeh, M.; Khoobi, M. A review: Biologically active 3,4-heterocycle-fused coumarins. *Eur. J. Med. Chem.* **2021**, *212*, 113034. [CrossRef]
8. Balalas, T.D.; Stratidis, G.; Papatheodorou, D.; Vlachou, E.-E.; Gabriel, C.; Hadjipavlou-Litina, D.J.; Litinas, K.E. One-pot Synthesis of 2-Substituted 4H-Chromeno[3,4-d]oxazol-4-ones from 4-Hydroxy-3-nitrocoumarin and Acids in the Presence of Triphenylphosphine and Phosphorus Pentoxide under Microwave Irradiation. *SynOpen* **2018**, *2*, 105–113. [CrossRef]
9. Prasanna, B.; Sandeep, A.; Revathi, T. Green approach to synthesis of novel substituted 8H-pyrano[2,3-e]benzoxazole-8-ones. *World J. Pharm. Pharm. Sci.* **2014**, *3*, 404–411.
10. Kontogiorgis, C.; Hadjipavlou-Litina, D. Biological Evaluation of Several Coumarin Derivatives Designed as Possible Anti-inflammatory/Antioxidant Agents. *J. Enzym. Inhib. Med. Chem.* **2003**, *18*, 63–69. [CrossRef]
11. Pathak, M.; Fellman, J.; Kaufman, K. The Effect of Structural Alterations on the Erythemal Activity of Furocoumarins: Psoralens ** From the Departments of Dermatology and Biochemistry, University of Oregon Medical School, Portland, Oregon and the Department of Chemistry, Kalamazoo College, Kalamazoo, Michigan. *J. Investig. Dermatol.* **1960**, *35*, 165–183. [CrossRef]
12. Soares, A.M.S.; Hungerford, G.; Gonçalves, M.S.T.; Costa, S.P.G. Light triggering of 5-aminolevulinic acid from fused coumarin ester cages. *New J. Chem.* **2017**, *41*, 2997–3005. [CrossRef]
13. Sahoo, S.S.; Shukla, S.; Nandy, S.; Sahoo, H. Synthesis of novel coumarin derivatives and its biological evaluations. *Eur. J. Exp. Biol.* **2012**, *2*, 899–908.
14. Colotta, V.; Catarzi, D.; Varano, F.; Cecchi, L.; Filacchioni, G.; Martini, C.; Giusti, L.; Lucacchini, A. Tricyclic heteroaromatic systems. Synthesis and benzodiazepine receptor affinity of 2-substituted-1-benzopyrano[3,4-d]oxazol-4-ones,-thiazol-4-ones, and -imidazol-4-ones. *Il Farm.* **1998**, *53*, 375–381. [CrossRef]
15. Nofal, Z.M.; El-Zahar, M.I.; El-Karim, S.S.A. Novel Coumarin Derivatives with Expected Biological Activity. *Molecules* **2000**, *5*, 99–113. [CrossRef]
16. Dallacker, F.; Kratzer, P.; Lipp, M. Derivate des 2.4-Pyronons und 4-Hydroxy-cumarins. *Eur. J. Org. Chem.* **1961**, *643*, 97–109. [CrossRef]

17. Gammon, D.W.; Hunter, R.; Wilson, S.A. An efficient synthesis of 7-hydroxy-2,6-dimethylchromeno[3,4-d]oxazol-4-one—A protected fragment of novenamine. *Tetrahedron* **2005**, *61*, 10683–10688. [CrossRef]

18. Saikachi, H.; Ichikawa, M. Studies on Synthesis of Coumarin Derivatives. XV. On the Preparation of Ethyl Pyranobenzoxazole-carboxylates. *Chem. Pharm. Bull.* **1966**, *14*, 1162–1167. [CrossRef] [PubMed]

19. Chantegrel, B.; Nadi, A.I.; Gelin, S. Synthesis of [1]benzopyrano[3,4-d]isoxazol-4-ones from 2-substituted chromone-3-carboxylic esters. A reinvestigation of the reaction of 3-acyl-4-hydroxycoumarins with hydroxylamine. Synthesis of 4-(2-hydroxybenzoyl)isoxazol-5-ones. *J. Org. Chem.* **1984**, *49*, 4419–4424. [CrossRef]

20. Kaufman, K.D.; McBride, D.W.; Eaton, D.C. Synthetic Furocoumarins. VII. Oxazolocoumarins from 6-Hydroxy-4-methylcoumarin. *J. Org. Chem.* **1965**, *30*, 4344–4346. [CrossRef]

21. Vlachou, E.-E.N.; Armatas, G.S.; Litinas, K.E. Synthesis of Fused Oxazolocoumarins from o -Hydroxynitrocoumarins and Benzyl Alcohol Under Gold Nanoparticles or FeCl3 Catalysis. *J. Heterocycl. Chem.* **2017**, *54*, 2447–2453. [CrossRef]

22. El-Saghier, A.M.M.; Naili, M.B.; Rammash, B.K.; Saleh, N.A.; Kreddan, K.M. Synthesis and antibacterial activity of some new fused chromenes. *Arkivoc* **2007**, *2007*, 83–91. [CrossRef]

23. Khan, I.A.; Kulkarni, M.V.; Gopal, M.; Shahabuddin, M.; Sun, C.-M. Synthesis and biological evaluation of novel angularly fused polycyclic coumarins. *Bioorg. Med. Chem. Lett.* **2005**, *15*, 3584–3587. [CrossRef] [PubMed]

24. Levrier, C.; Balastrier, M.; Beattie, K.D.; Carroll, A.; Martin, F.; Choomuenwai, V.; Davis, R.A. Pyridocoumarin, aristolactam and aporphine alkaloids from the Australian rainforest plant Goniothalamus australis. *Phytochemistry* **2013**, *86*, 121–126. [CrossRef] [PubMed]

25. Symeonidis, T.S.; Hadjipavlou-Litina, D.J.; Litinas, K.E. Synthesis Through Three-Component Reactions Catalyzed by FeCl3of Fused Pyridocoumarins as Inhibitors of Lipid Peroxidation. *J. Heterocycl. Chem.* **2013**, *51*, 642–647. [CrossRef]

26. Markey, M.D.; Fu, Y.; Kelly, T.R. Synthesis of Santiagonamine. *Org. Lett.* **2007**, *9*, 3255–3257. [CrossRef] [PubMed]

27. Kudale, A.A.; Kendall, J.; Miller, D.O.; Collins, J.L.; Bodwell, G.J. Povarov Reactions Involving 3-Aminocoumarins: Synthesis of 1,2,3,4-Tetrahydropyrido[2,3-c]coumarins and Pyrido[2,3-c]coumarins. *J. Org. Chem.* **2008**, *73*, 8437–8447. [CrossRef]

28. Heber, D.; Berghaus, T. Synthesis of 5H-[1]benzopyrano[4,3-b]pyridin-5-ones containing an azacannabinoidal structure. *J. Heterocycl. Chem.* **1994**, *31*, 1353–1359. [CrossRef]

29. Heber, D.; Ivanov, I.C.; Karagiosov, S.K. The vilsmeier reaction in the synthesis of 3-substituted [1]benzopyrano[4,3-b]pyridin-5-ones. An unusual pyridine ring closure. *J. Heterocycl. Chem.* **1995**, *32*, 505–509. [CrossRef]

30. Khan, A.T.; Das, D.K.; Islam, K.; Das, P. A simple and expedient synthesis of functionalized pyrido[2,3-c] coumarin derivatives using molecular iodine catalyzed three-component reaction. *Tetrahedron Lett.* **2012**, *53*, 6418–6422. [CrossRef]

31. Majumdar, K.; Ponra, S.; Ghosh, D.; Taher, A. Efficient One-Pot Synthesis of Substituted 4,7-Phenanthroline, Pyrano-[3,2-f]quinoline and Pyrano[3,2-g]quinoline Derivatives by Aza-Diels-Alder Reaction. *Synlett* **2010**, *2011*, 104–110. [CrossRef]

32. Han, Y.T.; Ahn, S.; Yoon, J.A. Total Synthesis of the Natural Pyridocoumarins Goniothaline A and B. *Synthesis* **2018**, *51*, 552–556. [CrossRef]

33. Symeonidis, T.S.; Kallitsakis, M.; Litinas, K.E. Synthesis of [5,6]-fused pyridocoumarins through aza-Claisen rearrangement of 6-propargylaminocoumarins. *Tetrahedron Lett.* **2011**, *52*, 5452–5455. [CrossRef]

34. Symeonidis, T.S.; Lykakis, I.N.; Litinas, K.E. Synthesis of quinolines and fused pyridocoumarins from N-propargylanilines or propargylaminocoumarins by catalysis with gold nanoparticles supported on TiO$_2$. *Tetrahedron* **2013**, *69*, 4612–4616. [CrossRef]

35. Ganguly, N.C.; Datta, M.; De, P.; Chakravarty, R. Studies on Regioselectivity of Nitration of Coumarins with Cerium(IV) Ammonium Nitrate: Solid-State Nitration of 6-Hydroxy-Coumarins on Montmorillonite K-10 Clay Support Under Microwave Irradiation. *Synth. Commun.* **2003**, *33*, 647–659. [CrossRef]

36. Lei, L.; Yang, D.; Liu, Z.; Wu, L. Mono-nitration of Coumarins by Nitric Oxide. *Synth. Commun.* **2004**, *34*, 985–992. [CrossRef]

37. De Araújo, R.S.A.; Guerra, F.Q.S.; Lima, E.D.O.; De Simone, C.A.; Tavares, J.F.; Scotti, L.; Scotti, M.T.; De Aquino, T.M.; De Moura, R.O.; Mendonça, F.J.B.; et al. Synthesis, Structure-Activity Relationships (SAR) and in Silico Studies of Coumarin Derivatives with Antifungal Activity. *Int. J. Mol. Sci.* **2013**, *14*, 1293–1309. [CrossRef]

38. Fountoulaki, S.; Daikopoulou, V.; Gkizis, P.L.; Tamiolakis, I.; Armatas, G.S.; Lykakis, I.N. Mechanistic Studies of the Reduction of Nitroarenes by NaBH4 or Hydrosilanes Catalyzed by Supported Gold Nanoparticles. *ACS Catal.* **2014**, *4*, 3504–3511. [CrossRef]

39. Wu, J.; Yu, J.; Wang, Y.; Zhang, P. Direct Amination of Phenols under Metal-Free Conditions. *Synlett* **2013**, *24*, 1448–1454. [CrossRef]

Communication

Synthesis of 4-[(1*H*-Benzimidazol-2-yl)sulfanyl]benzaldehyde and 2-({4-[(1*H*-Benzimidazol-2-yl)sulfanyl]phenyl}methylidene)hydrazine-1-carbothioamide

Mustafa Turki Ubeid, Hamdy Khamees Thabet * and Mohamed Yousef Abu Shuheil *

Chemistry Department, Faculty of Arts and Science, Northern Border University, P.O. Box 840, Rafha 91911, Saudi Arabia; mxt@gmail.com
* Correspondence: hamdy.salem@nbu.edu.sa (H.K.T.); mabushuheil@gmail.com (M.Y.A.S.);
  Tel.: +96-654-594-3150 (H.K.T.)

Abstract: Here we describe the preparation of 2-(4-((1*H*-benzo[d]imidazol-2-yl)thio)-benzylidene)-hydrazine-1-carbothioamide in two steps. In the first step, 1,3-dihydro-2H-1,3-benzimidazole-2-thione was reacted with 4-fluorobenzaldehyde in DMSO to get 4-[(1*H*-benzimidazol-2-yl)sulfanyl]benz aldehyde in high yield. The reaction of the obtained aldehyde with thiosemicarbazide in ethanol at reflux temperature yielded 2-({4-[(1*H*-benzimidazol-2-yl)sulfanyl]phenyl}methylidene)hydrazine-1-carbothioamide. The structure of the synthesized compounds was established by NMR spectroscopy (^{1}H, ^{13}C), mass spectrometry, and infrared spectroscopy.

Keywords: benzimidazole; nucleophilic aromatic substitution; thiosemicarbazone

Citation: Ubeid, M.T.; Thabet, H.K.; Abu Shuheil, M.Y. Synthesis of 4-[(1*H*-Benzimidazol-2-yl) sulfanyl]benzaldehyde and 2-({4-[(1*H*-Benzimidazol-2-yl) sulfanyl]phenyl}methylidene) hydrazine-1-carbothioamide. *Molbank* **2021**, *2021*, M1273. https://doi.org/10.3390/M1273

Academic Editors: Dimitrios Matiadis and Eleftherios Halevas

Received: 27 July 2021
Accepted: 18 August 2021
Published: 26 August 2021

Publisher's Note: MDPI stays neutral with regard to jurisdictional claims in published maps and institutional affiliations.

1. Introduction

Benzimidazoles are one of the important heterocyclic templates for the intensive investigation in chemical sciences due to their well-known applications and interesting biological activity profile [1–6]. Thiosemicarbazone moiety is another privileged structure that is found in several molecules with a wide range of biological activities representing several important classes in drug discovery [7–12]. Thiosemicarbazones have been postulated as biologically active compounds and display different types of biological activity, such as anticancer [11,13], anti-HIV [14,15], anticonvulsant [16,17], antimalarial [18,19], anti-inflammatory [20], enzymatic inhibition [9,10,21], antiviral [22], antioxidant [23], antifungal [24], and antibacterial [24,25]. Additionally, the flexibility of thiosemicarbazones as nitrogen and sulfur donors consents them to bring on a great diversity of coordination modes [26].

In this communication, we describe an improved process for the preparation of 4-[(1*H*-benzimidazol-2-yl)sulfanyl]benzaldehyde [27]. We also report the preparation of 2-({4-[(1*H*-benzimidazol-2-yl)sulfanyl]phenyl}methylidene)hydrazine-1-carbothioamide via the condensation of the synthesized aldehyde with thiosemicarbazide. The authors trust that this is the first report that discloses the synthesis and spectral analysis of 2-({4-[(1*H*-benzimidazol-2-yl)sulfanyl]phenyl}methylidene)hydrazine-1-carbothioamide because the exact structure search in the SciFinder database for this compound did not provide any hit or reference. The reported compounds can be assessed.

2. Results and Discussion

Refluxing of 1,3-dihydro-2*H*-1,3-benzimidazole-2-thione with 4-fluorobenzaldehyde in DMSO/anhydrous K_2CO_3 mixture gave 4-((1*H*-benzo[*H*]imidazol-2-yl)thio)benzaldehyde **3**. The product was isolated in 92% yield. The preparation of **3** is also provided in the literature [27], wherein a mixture of 1,3-dihydro-2*H*-1,3-benzimidazole-2-thione, 4-iodobenzaldehyde, CuI, 1,10-phenanthroline, K_2CO_3, and DMF was heated to 140 °C for

18 h. Further, the final product was purified by column chromatography to provide **3** in 87% yield. The current process does not make use of many reagents/parameters of the reported method, such as CuI, 1,10-phenanthroline, DMF, temperature, reaction duration, and column chromatography. These features make the current process an economical, safe, and less time-consuming process. Condensation of the latter with thiosemicarbazide in refluxing EtOH containing a catalytic amount of AcOH gave thiosemicarbazone derivative **4** (Scheme 1). The molecular structures of compounds **3** and **4** were established by spectral data. The IR spectrum of compound **3** displayed the presence of the imino group at 3153 cm^{-1}, carbonyl group at 1693 cm^{-1}, and C=N at 1590 cm^{-1}. (Figures S1 and S2) Additionally, compound **4** showed strong absorption bands at 3411, 3303, and 3158 cm^{-1} for the NH$_2$/NH groups and 1594 cm^{-1} for C=N. The ^{1}H NMR of **3** displayed the presence of singlet δ_H 12.52 for benzimidazole-NH and 9.95 for carboxaldehyde proton, and the aromatic protons (Ar-H) were found in the spectrum at δ_H 7.11–7.88. ^{13}C NMR (DMSO-d_6) showed signals at δ_C 192.4 assigned to the C=O group, 143.7 ppm assigned to C=N, 140.9 ppm assigned to C-N, in addition to 109.5 ppm assigned aromatic carbons at δ_C 134.7. The ^{1}H NMR of **4** exhibited the presence of an amino group at δ_H 8.01, azomethine proton at δ_H 8.24, a singlet at δ_H 11.49 for NH-CS, and a singlet at 12.84 for imidazole-NH, in addition to the presence of Ar-Hs at 7.19–8.46 ppm. ^{13}C NMR (DMSO-d_6) showed signals at δ_C 178.0 assigned to the C=S group, δ_C 145.9 assigned to CH=N, δ_C 143.7 assigned to C=N, δ_C 141.1 to C-N, and Ar-C at δ_C 135.3–111.1 (Figure S3–S6).

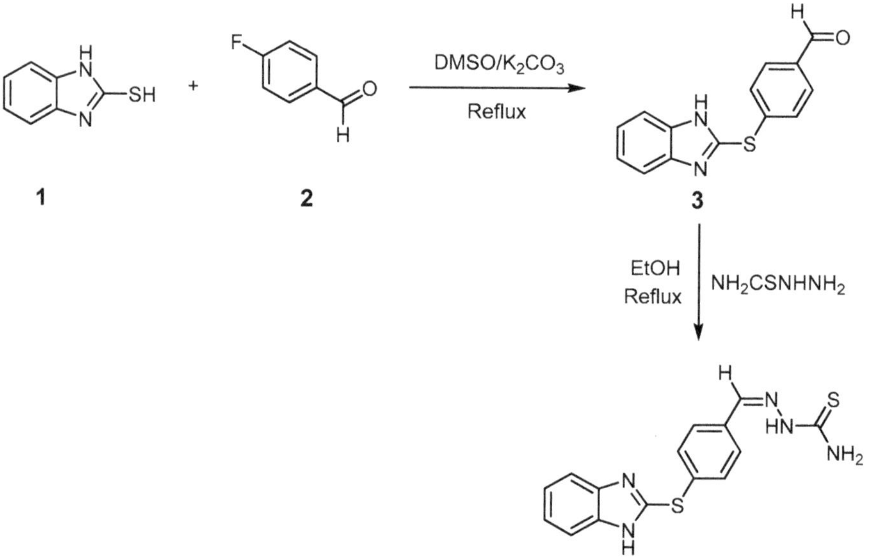

Scheme 1. Preparation of 4-[(1*H*-benzimidazol-2-yl)sulfanyl]benzaldehyde and 2-({4-[(1*H*-benzimidazol-2-yl)sulfanyl] phenyl}methylidene)hydrazine-1-carbothioamide.

3. Materials and Methods

3.1. General

The uncorrected melting points were determined by a Stuart melting point apparatus. IR (KBr) was obtained from a Shimadzu 440 spectrometer (ν, cm^{-1}). NMR spectra were recorded by a JEOL ECA-500 spectrometer. The chemical shifts (δ in ppm) were recorded relative to tetramethylsilane (TMS). The elemental analyses were performed at the Microanalytical Center, Cairo University, Cairo (Egypt).

3.2. 4-[(1H-Benzimidazol-2-yl)sulfanyl]benzaldehyde (3)

A mixture of 1,3-dihydro-2*H*-1,3-benzimidazole-2-thione 1 (10 mmol, 1.5 g) and 4-fluorobenzaldehyde 2 (10 mmol, 1.24 g) in dimethyl sulfoxide (25 mL) was refluxed along with anhydrous potassium carbonate (2 g) for 1 h, cooled, and transferred into crushed ice. The obtained product was collected and recrystallized to afford **3** as colorless solid. Yield: 2.34 g (92%); m.p.: 164–166 °C (acetic acid/water, 7:3); IR (KBr, cm^{-1}): 3153 (NH), 3069 (arom.-CH), 1693 (C=O), 1590 cm^{-1} (C=N); ^{1}H NMR (500 MHz, DMSO-d_6, δ/ppm): 7.10 (m, 1H, ArH), 7.20 (d, 2H, ArH, *J* = 5.0 Hz), 7.50 (dd, 3H, ArH, *J* = 5.0 Hz), 7.86 (d, 2H, ArH, *J* = 10.0 Hz), 9.95 (s, 1H, CHO), 12.52 (hump, 1H, benzimidazole-NH); ^{13}C NMR (125 MHz, DMSO-d_6): δ 109.5, 110.5, 119.4, 122.6, 124.3, 128.1, 128.9, 129.8, 130.2, 130.3, 130.8, 132.2, 134.7 (Ar'C), 140.9 (C-N), 143.7, 147.1 (C=N), 192.2 (C=O). Anal. calcd for C$_{14}$H$_{10}$N$_2$OS: C, 66.12; H, 3.96; N, 11.02; found: C, 65.94; H, 3.81; N, 10.87.

3.3. 2-({4-[(1H-Benzimidazol-2-yl)sulfanyl]phenyl}methylidene)hydrazine-1-carbothioamide (4)

A solution of aldehyde **3** (10 mmol, 2.54 g) and thiosemicarbazide (10 mmol, 0.91 g) was refluxed in EtOH (30 mL) containing acetic acid (5 mL) for 3 h. The obtained solid was collected and recrystallized to give **4**. Yellow crystals: yield: 2.45 g (75%); m.p., 268–270 °C (dioxane); IR (KBr, cm^{-1}): 3411, 3303, 3158 (NH$_2$/NH), 3014 (arom.-CH), 1594 (C=N); ^{1}H NMR (500 MHz, DMSO-d_6, δ/ppm): 7.15 (dd, 2H, ArH, *J* = 5.0 Hz), 7.41 (d, 1H, ArH, *J* = 5.0 Hz), 7.44 (d, 1H, ArH, *J* = 10.0 Hz), 7.57 (d, 1H, ArH, *J* = 5.0 Hz), 7.81 (d, 2H, ArH, *J* = 10.0 Hz), 8.01 (d, 2H, NH$_2$, *J* = 10.0 Hz), 8.24 (s, 1H, CH=N), 11.49 (s, 1H, NH-CS), 12.84 (s, 1H, benzimidazole-H); ^{13}C NMR (125 MHz, DMSO-d_6): δ 111.1, 118.4, 121.7, 122.8, 128.2, 130.7, 133.1, 133.9, 135.3 (Ar'C), 141.1 (C-N), 143.7 (C=N), 145.9 (CH=N), 178.0 (C=S). Anal. calcd for C$_{15}$H$_{13}$N$_5$S$_2$: C, 55.02; H, 4.00; N, 21.39; found: C, 54.86; H, 3.84; N, 21.23.

Supplementary Materials: The following are available online in Figures S1–S6 (FTIR, ^{1}H NMR, and ^{13}C NMR spectra of compounds **3** and **4**).

Author Contributions: Conceptualization, M.T.U., H.K.T. and M.Y.A.S.; methodology, M.T.U., H.K.T. and M.Y.A.S.; formal analysis, M.T.U., H.K.T. and M.Y.A.S.; writing—original draft preparation, M.T.U., H.K.T. and M.Y.A.S. All authors have read and agreed to the published version of the manuscript.

Funding: The research was carried out with the financial support of Northern Border University, Kingdom of Saudi Arabia in the framework of the scientific project 2017-1-8-F-7413.

Informed Consent Statement: Not applicable.

Data Availability Statement: Not applicable.

Acknowledgments: The authors would like to express their deep gratitude to Northern Border University, Kingdom of Saudi Arabia, for providing financial support for this research (project 2017-1-8-F-7535).

Conflicts of Interest: The authors declare no conflict of interest.

References

1. Wubulikasimu, R.; Yang, Y.; Xue, F.; Luo, X.; Shao, D.; Li, Y.; Ye, W. Synthesis and Biological Evaluation of Novel Benzimidazole Derivatives Bearing a Heterocyclic Ring at 4/5 Position. *Bull. Korean Chem. Soc.* **2013**, *34*, 2297–2304. [CrossRef]
2. Xiang, P.; Zhou, T.; Wang, L.; Sun, C.; Hu, J.; Zhao, Y.; Yang, L. Novel benzothiazole, benzimidazole and benzoxazole derivatives as potential antitumor agents: Synthesis and preliminary in vitro biological evaluation. *Molecules* **2012**, *17*, 873–883. [CrossRef]
3. Singh, M.; Tandon, V. Synthesis and biological activity of novel inhibitors of topoisomerase I: 2-Arylsubstituted 2-bis-1*H*-benzimidazoles. *Eur. J. Med. Chem.* **2011**, *46*, 659–669. [CrossRef] [PubMed]
4. Antoci, V.; Cucu, D.; Zbancioc, G.; Moldoveanu, C.; Mangalagiu, V.; Amariucai-Mantu, D.; Aricu, A.; Mangalagiu, I. Bis-(imidazole/benzimidazole)-pyridine derivatives: Synthesis, structure and antimycobacterial activity. *Future Med. Chem.* **2020**, *12*, 207–222. [CrossRef]
5. Gobis, K.; Foks, H.; Serocki, M.; Augustynowicz-kope, E.; Ave, G. Synthesis and evaluation of in vitro antimycobacterial activity of novel 1*H*-benzo[d]imidazole derivatives and analogues Agnieszka Napi o. *Eur. J. Med. Chem.* **2015**, *89*, 13–20. [CrossRef] [PubMed]

6. Awasthi, D.; Kumar, K.; Knudson, S.; Slayden, R.; Ojima, I. SAR studies on trisubstituted benzimidazoles as inhibitors of Mtb FtsZ for the development of novel antitubercular agents. *J. Med. Chem.* **2013**, *56*, 9756–9770. [CrossRef] [PubMed]

7. Liu, M.; Lin, T.; Cory, J.; Cory, A.; Sartorelli, A. Synthesis and biological activity of 3- and 5-amino derivatives of pyridine-2-carboxaldehyde thiosemicarbazone. *J. Med. Chem.* **1996**, *39*, 2586–2593. [CrossRef] [PubMed]

8. Tarasconi, P.; Capacchi, S.; Pelosi, G.; Cornia, M.; Albertini, R.; Bonati, A.; Dall'Aglio, P.; Lunghi, P.; Pinelli, S. Synthesis, spectroscopic characterization and biological properties of new natural aldehydes thiosemicarbazones. *Bioorg. Med. Chem.* **2000**, *8*, 157–162. [CrossRef]

9. Kumar, G.K.; Chavarria, G.E.; Charlton-Sevcik, A.K.; Arispe, W.M.; MacDonough, M.T.; Strecker, T.E.; Chen, S.E.; Siim, B.G.; Chaplin, D.J.; Trawick, M.L. Design, synthesis, and biological evaluation of potent thiosemicarbazone based cathepsin L inhibitors. *Bioorg. Med. Chem. Lett.* **2010**, *20*, 1415–1419. [CrossRef] [PubMed]

10. Fujii, N.; Mallari, J.P.; Hansell, E.J.; Mackey, Z.; Doyle, P.; Zhou, Y.; Gut, J.; Rosenthal, P.J.; McKerrow, J.H.; Guy, R.K. Discovery of potent thiosemicarbazone inhibitors of rhodesain and cruzain. *Bioorg. Med. Chem. Lett.* **2005**, *15*, 121–123. [CrossRef]

11. Hu, W.X.; Zhou, W.; Xia, C.N.; Wen, X. Synthesis and anticancer activity of Thiosemicarbazone. *Bioorg. Med. Chem. Lett.* **2006**, *16*, 2213–2218. [CrossRef]

12. Almutairi, M.S.; Zakaria, A.S.; Ignasius, P.P.; Al-Wabli, R.I.; Joe, I.H.; Attia, M.I. Synthesis, spectroscopic investigations, DFT studies, molecular docking and antimicrobial potential of certain new indole-isatin molecular hybrids: Experimental and theoretical approaches. *J. Mol. Struct.* **2018**, *1153*, 333–345. [CrossRef]

13. Bakherad, Z.; Safavi, M.; Fassihi, A.; Sadeghi-Aliabadi, H.; Bakherad, M.; Rastegar, H.; Ghasemi, J.B.; Sepehri, S.; Saghaie, L.; Mahdavi, M. Anti-cancer, anti-oxidant and molecular docking studies of thiosemicarbazone indole-based derivatives. *Res. Chem. Intermed.* **2019**, *45*, 2827–2854. [CrossRef]

14. Ishaq, M.; Taslimi, P.; Shafiq, Z.; Khan, S.; Salmas, R.E.; Zangeneh, M.M.; Saeed, A.; Zangeneh, A.; Sadeghian, N.; Asari, A.; et al. Synthesis, bioactivity and binding energy calculations of novel 3-ethoxysalicylaldehyde based thiosemicarbazone derivatives. *Bioorg. Chem.* **2020**, *100*, 103924. [CrossRef]

15. Hassan, M.; Ghaffari, R.; Sardari, S.; Farahani, Y.F.; Mohebbi, S. Discovery of novel isatin-based thiosemicarbazones: Synthesis, antibacterial, antifungal, and antimycobacterial screening. *Res. Pharm. Sci.* **2020**, *15*, 281–290. [PubMed]

16. Kshirsagar, A.; Toraskar, M.P.; Kulkarni, V.M.; Dhanashire, S.; Kadam, V. Microwave assisted synthesis of potential anti-infective and anticonvulsant thiosemicarbazones. *Int. J. Chem. Tech. Res.* **2009**, *1*, 696–701.

17. Matsa, R.; Makam, P.; Kaushik, M.; Hoti, S.L.; Kannan, T. Thiosemicarbazone derivatives: Design, synthesis and in vitro antimalarial activity studies. *Eur. J. Pharm. Sci.* **2019**, *137*, 104986. [CrossRef]

18. Oliveira, R.B.; Souza-Fagundes, E.M.; Soares, R.P.; Andrade, A.A.; Krettli, A.U.; Zani, C.L. Synthesis and antimalarial activity of semicarbazone and thiosemicarbazone derivatives. *Eur. J. Med. Chem.* **2008**, *43*, 1983–1988. [CrossRef]

19. Oliveira, J.; Nonato, F.; Zafred, R.; Leite, N.; Ruiz, A.; Carvalho, J.; Silva, A.; Moura, R.; Lim, M. Evaluation of anti-inflammatory effect of derivative (E)-N-(4-bromophenyl)-2-(thiophen- 2-ylmethylene)-thiosemicarbazone. *Biomed. Pharmacother.* **2016**, *80*, 388–392. [CrossRef]

20. Hałdys, K.; Latajka, R. Thiosemicarbazones with tyrosinase inhibitory activity. *Med. Chem. Commun.* **2019**, *10*, 378–389. [CrossRef]

21. Santacruz, M.; Fabiani, M.; Castro, E.; Cavallaro, L.; Finkielsztein, L. Synthesis, antiviral evaluation and molecular docking studies of N^4-aryl substituted/unsubstituted thiosemicarbazones derived from 1-indanones as potent anti-bovine viral diarrhea virus agents. *Bioorg. Med. Chem.* **2017**, *25*, 4055–4063. [CrossRef]

22. Nguyen, D.; Le, T.; Bui, T. Antioxidant activities of thiosemicarbazones from substituted benzaldehydes and N-(tetra-O-acetyl-β-D-galactopyranosyl)thiosemicarbazide. *Eur. J. Med. Chem.* **2013**, *60*, 199–207. [CrossRef] [PubMed]

23. Bartoli, J.; Montalbano, S.; Spadola, G.; Rogolino, D.; Pelosi, G.; Bisceglie, F.; Restivo, F.; Degola, F.; Serra, O.; Buschini, A.; et al. Antiaflatoxigenic Thiosemicarbazones as Crop-Protective Agents: A Cytotoxic and Genotoxic Study. *J. Agric. Food Chem.* **2019**, *67*, 10947–10953. [CrossRef] [PubMed]

24. Sens, L.; Souza, A.; Pacheco, L.; Menegatti, A.; Mori, M.; Mascarello, A.; Nunes, R.; Terenzi, H. Synthetic thiosemicarbazones as a new class of Mycobacterium tuberculosis protein tyrosine phosphatase A inhibitors. *Bioorg. Med. Chem.* **2018**, *26*, 5742–5750. [CrossRef] [PubMed]

25. Pedrido, R.; Gonzalez-Noya, A.M.; Romero, M.J.; Martinez-Calvo, M.; Lopez, M.V.; Gomez-Forneas, E.; Zaragoza, G.; Bermejo, M.R. Pentadentate thiosemicarbazones as versatile chelating systems. A comparative structural study of their metallic complexes. *Dalton Trans.* **2008**, *47*, 6776–6787. [CrossRef]

26. Yu, Y.; Kalinowski, D.S.; Kovacevic, Z.; Siafakas, A.R.; Jansson, P.J.; Stefani, C.; Lovejoy, D.B.; Sharpe, P.C.; Bernhardt, P.V.; Richardson, D.R. Thiosemicarbazones from the old to new: Iron chelators that are more than just ribonucleotide reductase inhibitors. *J. Med. Chem.* **2009**, *52*, 5271–5294. [CrossRef] [PubMed]

27. Koehler, A.N.; Stefan, E.; Caballero, F. Myc Modulators and Uses Thereof. U.S. Patent US10017520B2, 10 July 2018.

 *molbank*

Short Note

6-Bromo-*N*-(3-(difluoromethyl)phenyl)quinolin-4-amine

Christopher R. M. Asquith [1,2,*] and Graham J. Tizzard [3]

[1] Department of Pharmacology, University of North Carolina at Chapel Hill, Chapel Hill, NC 27599, USA
[2] Structural Genomics Consortium, UNC Eshelman School of Pharmacy, University of North Carolina at Chapel Hill, Chapel Hill, NC 27599, USA
[3] UK National Crystallography Service, School of Chemistry, Highfield Campus, University of Southampton, Southampton SO17 1BJ, UK; Graham.Tizzard@soton.ac.uk
* Correspondence: chris.asquith@unc.edu; Tel.: +1-919-491-3177

Received: 9 October 2020; Accepted: 15 October 2020; Published: 20 October 2020

Abstract: A routine synthesis was performed to furnish the title compound which incorporates a versatile difluoromethyl group on the aniline substitution of a 4-anilinoquinoline kinase inhibitor motif. In addition, the small molecule crystal structure (of the HCl salt) was solved, which uncovered that the difluoromethyl group was disordered within the packing arrangement and also a 126.08(7)° out of plane character between the respective ring systems within the molecule. The compound was fully characterized with ^{1}H/^{13}C-NMR and high-resolution mass spectra (HRMS), with the procedures described.

Keywords: 4-anilino-quin(az)olines; hinge binder; conformational flexibility; kinase inhibitor design

1. Introduction

The human kinome has shown impressive tractability, affording a number of high-profile clinical targets, which has led to the approval of more than 60 kinase inhibitors to date [1–3]. These drugs are for the most part multi-targeted tyrosine kinase inhibitors for the treatment of cancer [1–3]. There is an urgent need to develop highly selective compounds to open up the field into new indications beyond oncology [4]. However, this is not a straightforward task as all kinases bind a common substrate, adenosine triphosphate (ATP), which leads to a high degree of sequence homology across the kinome [5]. The 4-anilino-quin(az)oline scaffold shown is one of a number of hinge binders that have been shown to modulate kinome promiscuity, from broad spectrum such as GW559768X (**1**) to clinical narrow spectrum including lapatinib (**2**) and erlotinib and highly selective probes SGC-GAK-1 (**3**) (Figure 1) [6–13]. This kinome profile appears to be primarily driven by the electronics and substitution patterns of the pendant arms of the hinge binding scaffold [6–13].

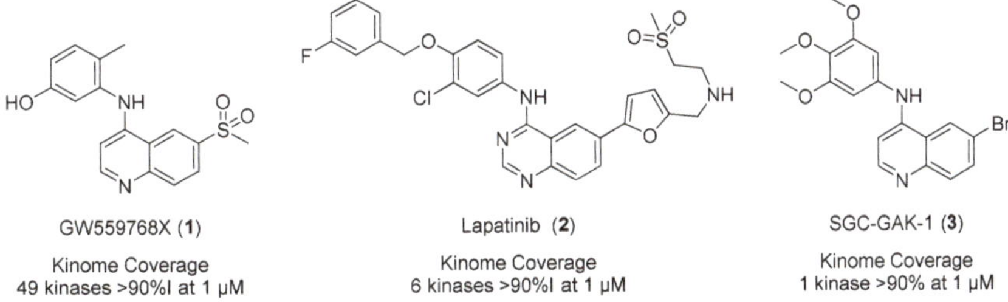

Figure 1. Previously reported 4-anilino-quin(az)olines (**1–3**) kinase inhibitors.

The difluoromethyl has shown a versatility in kinase inhibitor design, allowing the careful calibration of the compound properties [14–18]. These include potency profile, solubility and metabolic stability among others, several literature examples include PQR530 [19] and GDC-0077 [20,21] (Figure 2). We now describe incorporation of a difluoromethyl onto a known kinase active 6-bromoquinoline scaffold [6–10,22,23].

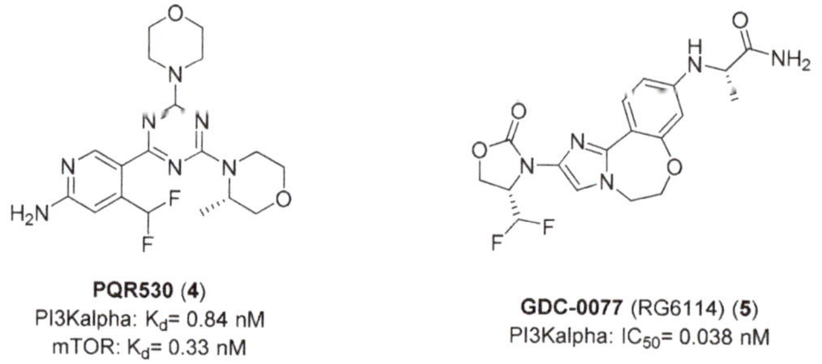

PQR530 (4)
PI3Kalpha: K_d= 0.84 nM
mTOR: K_d= 0.33 nM

GDC-0077 (RG6114) **(5)**
PI3Kalpha: IC_{50}= 0.038 nM

Figure 2. Examples of difluoromethyl use in kinase inhibitors (**4,5**).

2. Results

2.1. Synthesis of **8**

The title compound was synthesized by a one-step protocol (Scheme 1) [22–29]. The corresponding 3-(difluoromethyl)aniline (**6**) and 6-bromo-4-chloroquinoline (**7**) were mixed and refluxed for 16 h followed by the addition of Hünig's base. The resulting mixture was purified to afford the title compound (**8**) in very good overall yield (83%) (see supporting information).

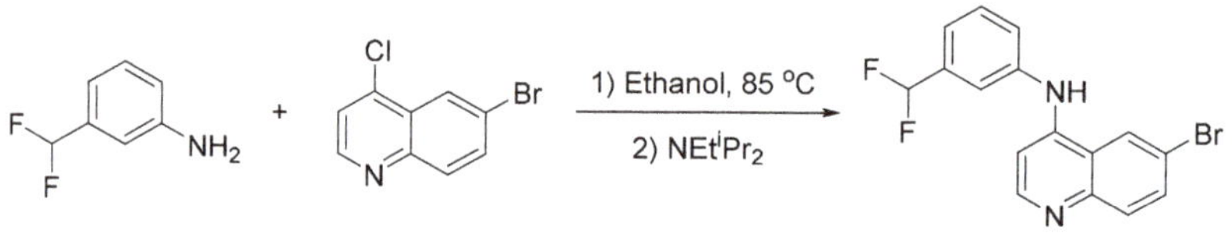

Scheme 1. Synthesis route to furnish 4-anilinoquinoline (**8**).

2.2. Crystal Structure of **8**

A crystallographic analysis revealed **8** crystallized as the chloride salt with the difluoromethyl group disordered over two positions with approximate proportions of 70:30 throughout the crystal (Figure 3). The molecule comprises two planar moieties. The quinoline moiety, i.e., N1 and C1–C9 exhibits a root mean square (r.m.s.) deviation of 0.006 Å with the maximum deviation from the plane being −0.011 Å for C2. The r.m.s. deviation of the difluoromethylphenyl moiety, i.e., C10–C15, is 0.009 Å with C12 and C15 displaying the maximum deviation of −0.012 Å and −0.013 Å, respectively. The dihedral angle between the two aforementioned planes is 126.08 (7)°. The C3–N2 bond distance of 1.342(3) Å is indicative of a double bond character in this bond consistent with conjugation in the quinoline moiety. All other bond distances and angles are within expected limits.

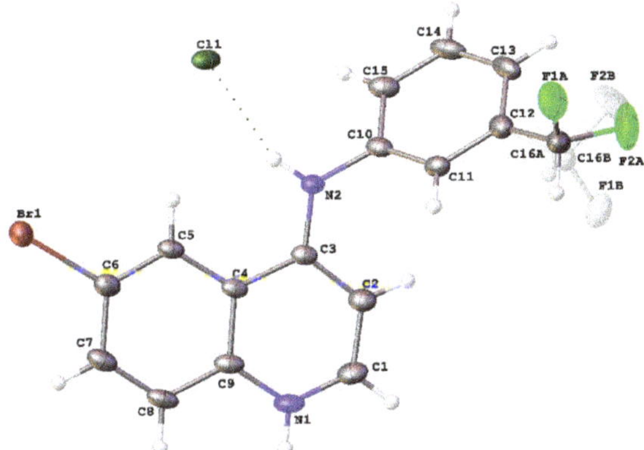

Figure 3. Molecular structure of 8H$^+$Cl$^-$ showing atomic labelling and displacement ellipsoids at 50% probability level. The difluromethyl minor disorder component is shown "ghosted".

The chloride anion is integral to the solid-state structure. The structure comprises corrugated "tapes" of 8H$^+$ hydrogen bonded to Cl$^-$ via the quinoline N–H$^+$ and aniline N–H, respectively (N–H$^+$ $\cdots$ Cl$^-$ $\cdots$ H–N (3.020(2) Å, 3.1113(19) Å)), parallel to the *b*-axis. These tapes form antiparallel stacks along the *a*-axis and close-pack along the *c*-axis to form the structure (Figure 4).

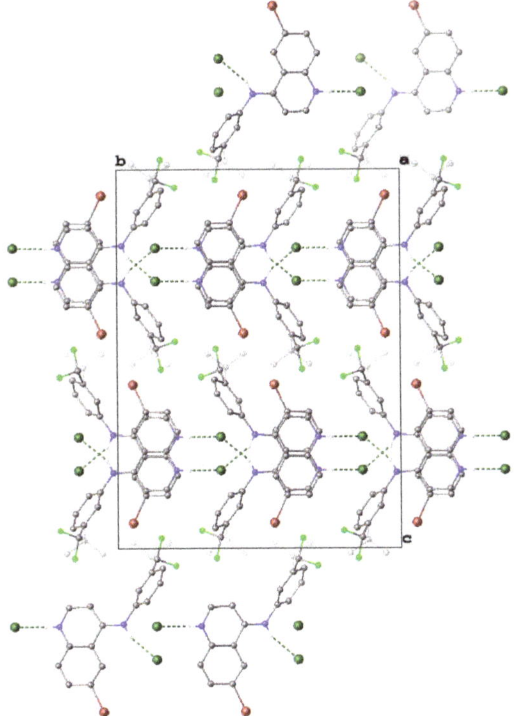

Figure 4. Unit cell contents of 8H$^+$Cl$^-$ shown in projection down the *a*-axis. Hydrogen atoms except those involved in H-bonding are omitted for clarity. Hydrogen bonds are shown as dashed green lines. Minor disorder components are shown "ghosted".

3. Discussion

We demonstrated a robust route to access the title compound (**8**) in excellent yield. This methodology lends itself to rapid library development as previously described [6–10,22–29]. The diversity of commercial and easy to synthetically access heterocycles and substitution patterns provide an endless wealth of possibilities and expandable chemical space to tune the 4-anilino-quin(az)oline scaffold properties for the discovery of new chemical tools and probes.

4. Material and Methods

1.1. Chemistry

All reactions were performed using flame-dried round-bottomed flasks or reaction vessels. Where appropriate, reactions were carried out under an inert atmosphere of nitrogen with dry solvents, unless otherwise stated. Yields refer to chromatographically and spectroscopically pure isolated yields. Reagents were purchased at the highest commercial quality and used without further purification. Reactions were monitored by thin-layer chromatography carried out on 0.25 mm E. Merck silica gel plates (60_{F-254}) using ultraviolet light as visualizing agent. NMR spectra were recorded on a Varian Inova 400 (Varian, Palo Alto, CA, USA) and were calibrated using residual undeuterated solvent as an internal reference (CDCl$_3$: ^{1}H-NMR = 7.26, ^{13}C-NMR = 77.16). The following abbreviations or combinations thereof were used to explain the multiplicities observed: s = singlet, d = doublet, t = triplet, q = quartet, m = multiplet, br = broad. Liquid chromatography (LC) and high-resolution mass spectra (HRMS) were recorded on a ThermoFisher hybrid LTQ FT (ICR 7T) (ThermoFisher, Waltham, MA, USA). The University of Southampton (Southampton, UK) small molecule x-ray facility collected and analyzed all X-ray diffraction data (Supplementary Materials).

6-Bromo-N-(3-(difluoromethyl)phenyl)quinolin-4-amine (**8**), 6-bromo-4-chloroquinoline (150 mg, 0.62 mmol, 1 eq) and 3-(difluoromethyl)aniline (97.3 mg, 0.68 mmol, 1.1 eq) were suspended in ethanol (10 mL) and refluxed for 18 h. This was followed by the addition of ^{i}Pr$_2$NEt (0.225 mL, 1.36 mmol, 2.2 eq.). The reaction mixture was concentrated in vacuo and extracted with ethyl acetate/saturated ammonium chloride. The title compound was then purified by flash chromatography using EtOAc: hexane followed by 1–5% methanol in EtOAc. The solvent was concentrated in vacuo and the product was obtained as a light beige solid (179 mg, 0.5134 mmol, 83%). MP 272–274 °C; ^{1}H-NMR (400 MHz, DMSO-d_6) δ 11.30 (s, 1H), 9.23 (d, J = 1.9 Hz, 1H), 8.56 (d, J = 6.9 Hz, 1H), 8.17 (dd, J = 9.0, 1.9 Hz, 1H), 8.12 (d, J = 9.0 Hz, 1H), 7.92–7.64 (m, 3H), 7.61 (dt, J = 6.4, 1.7 Hz, 1H), 7.14 (t, J = 55.7 Hz, 1H), 6.88 (d, J = 6.9 Hz, 1H). ^{13}C-NMR (100 MHz, DMSO-d_6) δ 153.9, 143.2, 137.7, 137.4, 136.6, 135.8 (t, J = 22.5 Hz), 130.6, 127.6 (t, J = 2.1 Hz), 126.4, 124.5 (t, J = 6.0 Hz), 123.0–121.8 (2C, m), 120.0, 118.8, 114.3 (t, J = 236.7 Hz), 100.5. HRMS *m/z* [M + H]$^+$ calcd for C$_{16}$H$_{12}$N$_2$BrF$_2$: 349.0152, found 349.0140, LC t_R = 3.78 min, >99% Purity.

4.2. Crystallography

Single colorless plate-shaped crystals of 8H$^+$Cl$^-$ were crystallized from EtOH/water and several drops of dioxane/HCL (4 M). A suitable crystal 0.16 × 0.10 × 0.02 mm^3 was selected and mounted on a MITIGEN holder (MiTeGen, Ithaca, NY, USA) in perfluoroether oil on a Rigaku FRE+ equipped with VHF Varimax confocal mirrors and an AFC12 goniometer and HyPix 6000 detector. The crystal was kept at a steady T = 100(2) K during data collection. The structure was solved with the ShelXT [30] structure solution program using the using dual methods solution method and by using Olex2 [31] as the graphical interface. The model was refined with version 2018/3 of ShelXL [32] using full matrix least squares minimization on F^2 minimization. All non-hydrogen atoms were refined anisotropically. Hydrogen atom positions were calculated geometrically and refined using the riding model except for those bonded to N-atoms which were located in the difference map and refined with a riding model. The difluoromethyl group was disordered over two positions (ca. 70:30). Thermal restraints and 1,2 and 1,3 equal distance geometric restraints applied to all equivalent atom pairs of disorder components. Equal distance geometric restraints were additionally applied to all C–F bonds.

Crystal Data for $C_{16}H_{12}N_2BrF_2Cl$ ($8H^+Cl^-$), M_r = 385.64, orthorhombic, *Pbca* (No. 61), a = 7.54250(10) Å, b = 17.2573(4) Å, c = 23.3121(7) Å, $a = b = g = 90°$, V = 3034.38(12) Å^3, T = 100(2) K, Z = 8, Z' = 1, m(Mo K$_a$) = 2.903 mm^{-1}, 26,395 reflections measured, 5342 unique (R_{int} = 0.0384) which were used in all calculations. The final wR_2 was 0.0990 (all data) and R_1 was 0.0539 (I ≥ 2 s(I)).

Supplementary Materials: The following are available online, ^{1}H and ^{13}C NMR spectra and the crystallographic data for Compound $8H^+Cl^-$ in crystallographic information file (CIF) format. CCDC 2036373 also contains the supplementary crystallographic data for this paper. These data can be obtained free of charge via http://www.ccdc.cam.ac.uk/conts/retrieving.html.

Author Contributions: Conceptualization and writing the original draft preparation: C.R.M.A.; validation, resources, data curation and editing: C.R.M.A. and G.J.T. All authors have read and agreed to the published version of the manuscript.

Funding: The SGC is a registered charity (number 1097737) that receives funds from AbbVie, Bayer Pharma AG, Boehringer Ingelheim, Canada Foundation for Innovation, Eshelman Institute for Innovation, Genome Canada, Innovative Medicines Initiative (EU/EFPIA) (ULTRA-DD grant no. 115766), Janssen, Merck KGaA Darmstadt Germany, MSD, Novartis Pharma AG, Ontario Ministry of Economic Development and Innovation, Pfizer, São Paulo Research Foundation-FAPESP, Takeda, and Wellcome (106169/ZZ14/Z).

Acknowledgments: We thank the EPSRC UK National Crystallography Service at the University of Southampton for the collection of the crystallographic data. We also thank Brandie Ehrmann for LC–MS/HRMS support provided by the Mass Spectrometry Core Laboratory at the University of North Carolina at Chapel Hill.

Conflicts of Interest: The authors declare no conflict of interest.

References

1. Ferguson, F.M.; Gray, N.S. Kinase inhibitors: The road ahead. *Nat. Rev. Drug Discov.* **2018**, *17*, 353–377. [CrossRef] [PubMed]

2. Roskoski, R., Jr. Properties of FDA-approved small molecule protein kinase inhibitors. *Pharmacol. Res.* **2019**, *144*, 19–50. [CrossRef] [PubMed]

3. Roskoski, R., Jr. Properties of FDA-approved small molecule protein kinase inhibitors: A 2020 update. *Pharmacol. Res.* **2020**, *152*, 104609. [CrossRef] [PubMed]

4. Cohen, P.; Alessi, D.R. Kinase drug discovery—What's next in the field? *ACS Chem. Biol.* **2013**, *8*, 96–104. [CrossRef]

5. Manning, G.; Whyte, D.B.; Martinez, R.; Hunter, T.; Sudarsanam, S. The protein kinase complement of the human genome. *Science* **2002**, *298*, 1912–1934. [CrossRef]

6. Asquith, C.R.M.; Laitinen, T.; Bennett, J.M.; Godoi, P.H.; East, M.P.; Tizzard, G.J.; Graves, L.M.; Johnson, G.L.; Dornsife, R.E.; Wells, C.I.; et al. Identification and Optimization of 4-Anilinoquinolines as Inhibitors of Cyclin G Associated Kinase. *ChemMedChem* **2018**, *13*, 48–66. [CrossRef]

7. Asquith, C.R.M.; Berger, B.T.; Wan, J.; Bennett, J.M.; Capuzzi, S.J.; Crona, D.J.; Drewry, D.H.; East, M.P.; Elkins, J.M.; Fedorov, O.; et al. SGC-GAK-1: A Chemical Probe for Cyclin G Associated Kinase (GAK). *J. Med. Chem.* **2019**, *62*, 2830–2836. [CrossRef]

8. Asquith, C.R.M.; Naegeli, N.; East, M.P.; Laitinen, T.; Havener, T.M.; Wells, C.I.; Johnson, G.L.; Drewry, D.H.; Zuercher, W.J.; Morris, D.C. Design of a cyclin G associated kinase (GAK)/epidermal growth factor receptor (EGFR) inhibitor set to interrogate the relationship of EGFR and GAK in chordoma *J. Med. Chem.* **2019**, *62*, 4772–4778. [CrossRef]

9. Asquith, C.R.M.; Treiber, D.K.; Zuercher, W.J. Utilizing comprehensive and mini-kinome panels to optimize the selectivity of quinoline inhibitors for cyclin G associated kinase (GAK). *Bioorg. Med. Chem. Lett.* **2019**, *29*, 1727–1731. [CrossRef]

10. Asquith, C.R.M.; Bennett, J.M.; Su, L.; Laitinen, T.; Elkins, J.M.; Pickett, J.E.; Wells, C.I.; Li, Z.; Willson, T.M.; Zuercher, W.J. Development of SGC-GAK-1 as an orally active in vivo probe for cyclin G associated kinase through cytochrome P450 inhibition. *bioRxiv* **2019**, 629220. [CrossRef]

11. Drewry, D.H.; Wells, C.I.; Andrews, D.M.; Angell, R.; Al-Ali, H.; Axtman, A.D.; Capuzzi, S.J.; Elkins, J.M.; Ettmayer, P.; Frederiksen, M.; et al. Progress towards a public chemogenomic set for protein kinases and a call for contributions. *PLoS ONE* **2017**, *12*, e0181585. [CrossRef] [PubMed]

12. Fabian, M.A.; Biggs, W.H., III; Treiber, D.K.; Atteridge, C.E.; Azimioara, M.D.; Benedetti, M.G.; Carter, T.A.; Ciceri, P.; Edeen, P.T.; Floyd, M.; et al. A small molecule-kinase interaction map for clinical kinase inhibitors. *Nat. Biotechnol.* **2005**, *23*, 329–336. [CrossRef] [PubMed]

13. Karaman, M.W.; Herrgard, S.; Treiber, D.K.; Gallant, P.; Atteridge, C.E.; Campbell, B.T.; Chan, K.W.; Ciceri, P.; Davis, M.I.; Edeen, P.T.; et al. A quantitative analysis of kinase inhibitor selectivity. *Nat. Biotechnol.* **2008**, *26*, 127–132. [CrossRef] [PubMed]

14. Shah, P.; Westwell, A.D. The role of fluorine in medicinal chemistry. *J. Enzyme Inhib. Med. Chem.* **2007**, *22*, 527–540. [CrossRef]

15. Hagmann, W.K. The Many Roles for Fluorine in Medicinal Chemistry. *J. Med. Chem.* **2008**, *51*, 4359–4369. [CrossRef]

16. Swallow, S. Chapter Two-Fluorine in Medicinal Chemistry. *Prog. Med. Chem.* **2015**, *54*, 65–133. [CrossRef]

17. Wang, Y.; Callejo, R.; Slawin, A.M.Z.; O'Hagan, D. The difluoromethylene (CF2) group in aliphatic chains: Synthesis and conformational preference of palmitic acids and nonadecane containing CF2 groups Beilstein. *J. Org. Chem.* **2014**, *10*, 18–25. [CrossRef]

18. Corr, M.J.; Cormanich, R.A.; von Hahmann, C.N.; Bühl, M.; Cordes, D.B.; Slawin, A.M.Z.; O'Hagan, D. Fluorine in fragrances: Exploring the difluoromethylene (CF2) group as a conformational constraint in macrocyclic musk lactones. *Org. Biomol. Chem.* **2016**, *14*, 211. [CrossRef]

19. Rageot, D.; Bohnacker, T.; Keles, E.; McPhail, J.A.; Hoffmann, R.M.; Melone, A.; Borsari, C.; Sriramaratnam, R.; Sele, A.M.; Beaufils, F.; et al. (S)-4-(Difluoromethyl)-5-(4-(3-methylmorpholino)-6-morpholino-1,3,5-triazin-2-yl)pyridin-2-amine (PQR530), a potent, orally bioavailable, and brain-penetrable dual inhibitor of class I PI3K and mTOR kinase. *J. Med. Chem.* **2019**, *62*, 6241–6261. [CrossRef]

20. Braun, M.-G.; Hanan, E.; Staben, S.T.; Heald, R.A.; MacLeod, C.; Elliott, R. Benzoxazepin Oxazolidinone Compounds and Methods of Use. U.S. Patent Application 20,170,210,733, 16 May 2017.

21. Han, C.; Kelly, S.M.; Cravillion, T.; Savage, S.J.; Nguyen, T.; Gosselin, F. Synthesis of PI3K inhibitor GDC-0077 via a stereocontrolled *N*-arylation of α-amino acids. *Tetrahedron* **2019**, *75*, 4351–4357. [CrossRef]

22. Asquith, C.R.M.; Tizzard, G.J.; Bennett, J.M.; Wells, C.I.; Elkins, J.M.; Willson, T.; Poso, A.; Laitinen, T. Targeting the water network in cyclin G associated kinase (GAK) with 4-anilino-quin(az)oline inhibitors. *ChemMedChem* **2020**, *15*, 1200–1215. [CrossRef]

23. Asquith, C.R.M.; Laitinen, T.; Bennett, J.M.; Wells, C.I.; Elkins, J.M.; Zuercher, W.J.; Tizzard, G.J.; Poso, A. Design and analysis of the 4-anilino-quin(az)oline kinase inhibition profiles of GAK/SLK/STK10 using quantitative structure activity relationships. *ChemMedChem* **2020**, *15*, 26–49. [CrossRef] [PubMed]

24. Saul, S.; Pu, S.; Zuercher, W.J.; Einav, S.; Asquith, C.R.M. Potent antiviral activity of novel multi-substituted 4-anilinoquin(az)olines. *Bioorg. Med. Chem. Lett.* **2020**, *30*, 127284. [CrossRef] [PubMed]

25. Asquith, C.R.M.; Laitinen, T.; Wells, C.I.; Tizzard, G.J.; Zuercher, W.J. New insights into 4-anilinoquinazolines as inhibitors of cardiac troponin I–interacting kinase (TNNi3K). *Molecules* **2020**, *25*, 1697. [CrossRef]

26. Asquith, C.R.M.; Tizzard, G.J. 6-Bromo-N-(2-methyl-2H-benzo[d][1,2,3]triazol-5-yl)quinolin-4-amine. *Molbank* **2019**, *2019*, 1087. [CrossRef]

27. Carabajal, M.A.; Asquith, C.R.M.; Laitinen, T.; Tizzard, G.J.; Yim, L.; Rial, A.; Chabalgoity, J.; Zuercher, W.J.; Véscovi, E.G. Quinazoline-based anti-virulence compounds that selectively target Salmonella PhoP/PhoQ signal transduction system. *Antimicrob. Agents Chemother.* **2019**, *64*, e01744-19. [CrossRef] [PubMed]

28. Asquith, C.R.M.; Fleck, N.; Torrice, C.D.; Crona, D.J.; Grundner, C.; Zuercher, W.J. Anti-tubercular activity of novel 4-anilinoquinolines and 4-anilinoquinazolines. *Bioorg. Med. Chem. Lett.* **2019**, *18*, 2695–2699. [CrossRef] [PubMed]

29. Asquith, C.R.M.; Maffuid, K.A.; Laitinen, T.; Torrice, C.D.; Tizzard, G.J.; Crona, D.J.; Zuercher, W.J. Targeting an EGFR water network using novel 4-anilinoquin(az)olines inhibitors for chordoma. *ChemMedChem* **2019**. [CrossRef]

30. Sheldrick, G.M. ShelXT-Integrated space-group and crystal-structure determination. *Acta Cryst.* **2015**, *A71*, 3–8. [CrossRef]

31. Dolomanov, O.V.; Bourhis, L.J.; Gildea, R.J.; Howard, J.A.K.; Puschmann, H. Olex2: A complete structure solution, refinement and analysis program. *J. Appl. Cryst.* **2009**, *42*, 339–341. [CrossRef]
32. Sheldrick, G.M. Crystal structure refinement with ShelXL. *Acta Cryst.* **2015**, *C27*, 3–8. [CrossRef]

Publisher's Note: MDPI stays neutral with regard to jurisdictional claims in published maps and institutional affiliations.

Short Note

4′-(5-*N*-Propylthiophen-2-yl)-2,2′:6′,2″-terpyridine

Jérôme Husson * and Laurent Guyard

Institut Utinam UMR CNRS 6213, UFR Sciences et Techniques, Université de Bourgogne-Franche-Comté, 16 route de Gray, 25030 Besançon, France; laurent.guyard@univ-fcomte.fr
* Correspondence: jerome.husson@univ-fcomte.fr; Tel.: +33-3-81666291

Abstract: A new thiophene-substituted terpyridine derivative has been prepared through the reaction between 5-*N*-propylthiophene-2-carboxaldehyde and 2-acetylpyridine. This terpyridine derivative bears an alkyl chain linked via a thiophene heterocycle.

Keywords: chalcogenophene; heterocycles; ligands; pyridine derivatives; thiophene derivatives

Citation: Husson, Jrm.; Guyard, L. 4′-(5-*N*-Propylthiophen-2-yl)-2,2′:6′,2″-terpyridine. *Molbank* **2021**, *2021*, M1183. https://doi.org/10.3390/M1183

Academic Editors: Dimitrios Matiadis and Eleftherios Halevas

Received: 8 December 2020
Accepted: 20 January 2021
Published: 21 January 2021

Publisher's Note: MDPI stays neutral with regard to jurisdictional claims in published maps and institutional affiliations.

1. Introduction

2,2′:6′,2″-Terpyridine (terpy) ligands and their metal complexes have been widely studied [1] owing to the broad range of applications for such molecules. Varying the nature of the substituents on the ligands and/or the metallic centre offers the possibility to prepare an enormous number of different substances. In particular, terpyridines that contain the five membered heterocycle thiophene [2] have attracted widespread of attention. In fact, they can be used in the preparation of materials for solar cells [3–5], for the functionalization of nanoparticles [6], as fluorescent probes [7,8], as antimicrobial agents [9], as electrochromic materials [10] or as chromophores [11], just to name a few. The substituents that are present on the thiophene ring have an important impact, especially on properties of thiophene-substituted terpyridine-based materials [12]. Therefore, the preparation of new thiophene-substituted terpyridines is still of interest. This paper presents the synthesis of the novel 4′-(5-*N*-propylthiophen-2-yl)-2,2′:6′,2″-terpyridine ligand (**1**) (Figure 1).

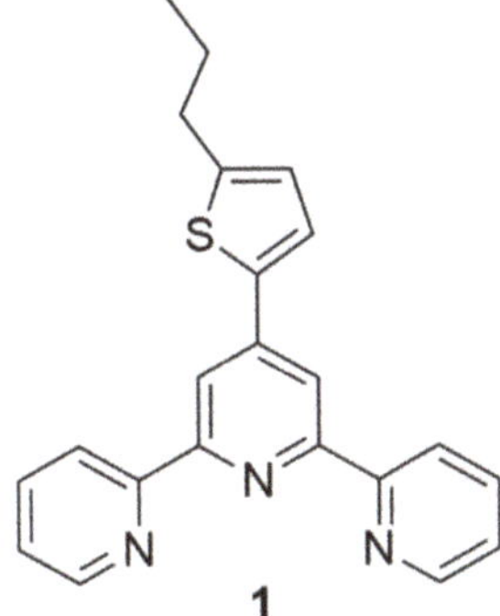

Figure 1. Chemical structure 4′-(5-*N*-propylthiophen-2-yl)-2,2′:6′,2″-terpyridine (**1**).

2. Results and Discussion

Many synthetic methods are available for the preparation of terpyridine derivatives [13–15]. In order to prepare 4′-(5-*N*-propylthiophen-2-yl)-2,2′:6′,2″-terpyridine, the method described by Wang and Hanan in 2005 [16] was selected. This procedure allowed the facile preparation of **1** from 2-acetylpyridine and 5-*N*-propylthiophene-2-carboxaldehyde.

As in many cases with this synthetic protocol, the crude product was sufficiently pure (>98% by quantitative NMR [17,18] and by combustion analysis) to be used (e.g., for the preparation of metal complexes) without purification.

Ligand **1** was characterized by ^{1}H and ^{13}C-NMR as well as by HR-MS. Firstly, the ^{1}H-NMR spectrum agrees with the chemical structure. NMR spectra of 4'-functionalized terpyridines exhibit a typical singlet for proton 3' and 5'. In the present molecule, this singlet is seen at δ = 8.63 ppm. Furthermore, as expected, hydrogens that belong to the thiophene heterocycle (a and b) appear as doublets centered at 7.60 and 6.84 ppm, respectively, with a coupling constant of 3.6 Hz. Finally, signals for the propyl chain can be observed as two triplets (at 2.84 and 1.02 ppm) and a multiplet at 1.76 ppm (Figure 2).

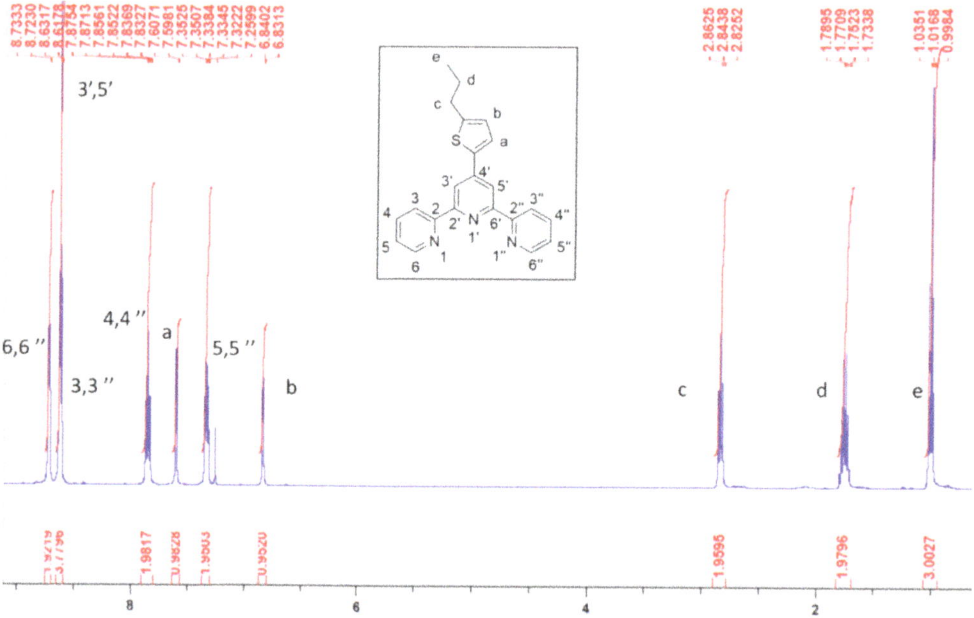

Figure 2. ^{1}H-NMR spectra of compound **1** (inset: structure and atom numbering of **1**).

Additionally, the structure of **1** was further confirmed by ^{13}C-NMR as well as by HR-MS (Supplementary Materials). For instance, the ^{13}C-NMR spectrum features 15 signals due to the symmetry of the molecule, while mass spectra exhibit the molecular ion peak at 358.13703 (calc. for $[C_{22}H_{19}N_3S + H]^+$: 358.13724). The UV-Vis spectrum of compound **1** recorded in acetonitrile exhibits bands at 231, 252, 286 and 314 nm (Figure 3).

The strong absorption band can be assigned to $\pi - \pi^*$ transitions of the terpyridine part, and the shape of the spectrum is similar to previously reported ones for such five-membered heterocycle-substituted terpyridines [11,19].

As expected, introduction of an aliphatic chain onto the thiophene ring subsequently lowered the melting point of the product (98–99 °C) when compared to other non-functionalized chalcogenophene-substituted terpyridine molecule [20–23]. This phenomenon is also observed in an hexylthiophene-functionalized 2,2':2',2"-terpyridine [11] (Table 1).

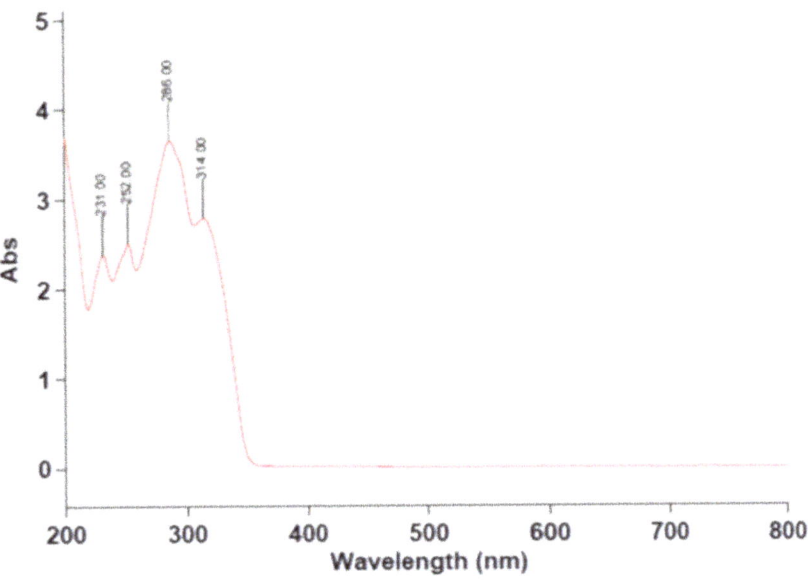

Figure 3. UV-Vis. spectrum of terpyridine **1** (1.21×10^{-4} M in acetonitrile).

Table 1. Comparison of melting points for some chalcogenophene-substituted terpyridines. [a] Values from literature.

X=	S			O	Se	Te
R=	$n\mathrm{C_6H_{13}}$	$n\mathrm{C_3H_7}$	H	H	H	H
Mp (°C)	70–72 [a]	98–99	197–199 [a]	219 [a]	215–218 [a]	196–200 [a]

3. Materials and Methods

All reagents were purchased from commercial suppliers (ACROS Organics, Geel, Belgium and TCI Chemicals, Zwijndrecht, Belgium) and used as received. Starting material 5-*N*-propylthiophene-2-carboxaldehyde [24] was prepared from thiophene via 2-*N*-propylthiophene [25] and 1-(thiophen-2-yl)-propan-1-one [26] according to literature procedures. ^{1}H and ^{13}C-NMR spectra were recorded on a Brucker AC 400 (Bruker, Wissembourg, France) at 400 and 100 MHz, respectively, using CDCl$_3$ as a solvent. Melting point was recorded with a Stuart SMP 10 melting point apparatus (Bibby Sterilin, Stone, UK) and was uncorrected. The UV-Vis spectrum was recorded on a Cary 300 (Agilent Technologies, Santa Clara, CA, USA) using acetonitrile (C = 1.21×10^{-4} M) as solvent. HR-MS was recorded at Sayence SATT, Dijon, France. Elemental analysis was performed at Service d'Analyse Élémentaire, Vandoeuvre-les-Nancy, France.

4'-(5-*N*-Propylthiophen-2-yl)-2,2':6',2"-terpyridine (**1**): to a solution of 2-acetylpyridine (7.43 g; 61 mmol) in ethanol (154 mL), 5-*N*-propylthiophene-2-carboxaldehyde (4.73 g; 31 mmol), 85% potassium hydroxide pellets (4.73 g; 72 mmol) and 25% aqueous ammonia (89 mL) were added. The reaction mixture was stirred at room temperature for 24 h. The solid was then filtered on a glass sintered funnel and washed with ice-cold 50% ethanol until washings were colorless. The product was dried under vacuum over phosphorus pentoxide. Compound **1** was obtained as a light-yellow solid (4.25 g; 39%). Mp= 98–99 °C. ^{1}H-NMR (CDCl$_3$, 400 MHz), δ (ppm): 8.73 (d, 2H H6, 6″, *J* = 4.2 Hz), 8.63 (d, 2H, H3, 3″, *J* = 7.6 Hz), 8.63 (s, 2H, H3′, 5′), 7.85 (m, 2H, H4, 4″), 7.60 (d, 1H, Ha, *J* = 3.6 Hz), 7.34 (dd, 2H, H5, 5″, *J* = 6.5 Hz, *J* = 4.9 Hz), 6.84 (d, 1H, Hb, *J* = 3.6 Hz),2.84 (t, 2H, Hc, *J* = 7.5 Hz), 1.76 (m, 2H, Hd), 1.02 (t, 3H, He, *J* = 7.3 Hz). ^{13}C-NMR (CDCl$_3$, 100 MHz), δ (ppm): 156.2, 155.9, 149.1, 148.1, 143.8, 139.1, 136.8, 125.7, 125.6, 123.8, 121.3, 116.7, 32.4, 24.8, 13.7. HR-MS: calc. for $[C_{22}H_{19}N_3S + H]^+$ 358.13724, found 358.13703. Elemental analysis for $C_{22}H_{19}N_3S$: C, 73.92; H, 5.36; N, 11.75; S, 8.97, found C, 73.27; H, 5.39; N, 11.90; S, 8.96. UV-Vis (nm): λ_{abs} = 231, 252, 286, 314.

4. Conclusions

A new thiophene-containing terpyridine was prepared and characterized. This ligand features an alkyl chain on the thiophene ring. This resulted in a lowering of the melting point of this type of molecule, a feature that can be interesting in view of future applications (e.g., for the preparation of low melting complexes).

Experiments are currently in progress to incorporate this ligand into organometallic materials. Results will be reported in due course.

Supplementary Materials: The following are available online, ^{1}H and ^{13}C-NMR, HR-MS, UV-Vis and IR spectra of terpyridine **1**.

Author Contributions: J.H. conceived and carried out the experiments, analyzed data and prepared the manuscript. L.G. analyzed data and contributed to manuscript preparation. All authors have read and agreed to the published version of the manuscript.

Funding: This research did not receive specific funding.

Data Availability Statement: The data from this study are available in this paper and in its Supplementary Materials.

Conflicts of Interest: The authors declare no conflict of interest.

References

1. Schubert, U.S.; Hofmeier, H.; Newkome, G.R. *Modern Terpyridine Chemistry*; Wiley-VCH: Weinheim, Germany, 2006.
2. Husson, J.; Knorr, M. 2,2′:6′,2″-Terpyridines Functionalized with Thienyl Substituents: Synthesis and Applications. *J. Heterocycl. Chem.* **2012**, *49*, 453–478. [CrossRef]
3. Vincent Joseph, K.L.; Anthonysamy, A.; Easwaramoorthi, R.; Shinde, D.V.; Ganapathy, V.; Karthikeyan, S.; Lee, J.; Park, T.; Rhee, S.-W.; Kim, K.S.; et al. Cyanoacetic acid tethered thiophene for well-matched LUMO level in Ru(II)-terpyridine dye sensitized solar cells. *Dyes Pigment.* **2016**, *126*, 270–278. [CrossRef]
4. Dehaudt, J.; Husson, J.; Guyard, L.; Oswald, F.; Martineau, D. A simple access to "Black-Dye" analogs with good efficiencies in dye-sensitized solar cells. *Renew. Energy* **2014**, *66*, 588–595. [CrossRef]
5. Caramori, S.; Husson, J.; Beley, M.; Bignozzi, C.A.; Argazzi, R.; Gros, P.C. Combination of Cobalt and Iron Polypyridine Complexes for Improving the Charge Separation and Collection in Ru(terpyridine)$_2$-Sensitised Solar Cells. *Chem. Eur. J.* **2010**, *16*, 2611–2618. [CrossRef] [PubMed]
6. Pruskova, M.; Sutrova, V.; Slouf, M.; Vlckova, B.; Vohlidal, J.; Sloufova, I. Arrays of Ag and Au Nanoparticles with Terpyridine- and Thiophene-Based Ligands: Morphology and Optical Responses. *Langmuir* **2017**, *33*, 4146–4156. [CrossRef]
7. Shen, Y.; Shao, T.; Fang, B.; Du, W.; Zhang, M.; Liu, J.; Liu, T.; Tian, X.; Zhang, Q.; Wang, A.; et al. Visualization of mitochondrial DNA in living cells with super-resolution microscopy using thiophene-based terpyridine Zn(II) complexes. *Chem. Commun.* **2018**, *54*, 11288–11291. [CrossRef]
8. Feng, Z.; Li, D.; Zhang, M.; Shao, T.; Shen, T.; Tian, X.; Zhang, Q.; Li, S.; Wu, J.; Tian, Y. Enhanced three-photon activity triggered by the AIE behavior of a novel terpyridine-based Zn(II) complex bearing a thiophene bridge. *Chem. Sci.* **2019**, *10*, 7228–7232. [CrossRef]
9. Njogu, E.M.; Martincigh, B.S.; Omondi, B.; Nyamori, V.O. Synthesis, characterization, antimicrobial screening and DNA binding of novel silver(I)-thienylterpyridine and silver(I)-furylterpyridine. *Appl. Organomet. Chem.* **2018**, *32*, e4554. [CrossRef]
10. Liang, Y.W.; Strohecker, D.; Lynch, V.; Holliday, B.J.; Jones, R.A. A Thiophene-Containing Conductive Metallopolymer Using an Fe(II) Bis(terpyridine) Core for Electrochromic Materials. *ACS Appl. Mater. Interfaces* **2016**, *8*, 34568–34580. [CrossRef]
11. Fernandes, S.S.M.; Besley, M.; Ciarrocchi, C.; Licchelli, M.; Raposo, M.M.M. Terpyridine derivatives functionalized with (hetero)aromatic groups and the corresponding Ru complexes: Synthesis and characterization as SHG chromophores. *Dyes Pigment.* **2018**, *150*, 49–58. [CrossRef]

12. Mukherjee, S.; Torres, D.E.; Jakubikova, E. HOMO inversion as a strategy for improving the light-absorption properties of Fe(II) chromophores. *Chem. Sci.* **2017**, *8*, 8115–8126. [CrossRef] [PubMed]

13. Heller, M.; Schubert, U.S. Syntheses of functionalized 2,2′:6′,2″-terpyridines. *Eur. J. Org. Chem.* **2003**, *6*, 947–961. [CrossRef]

14. Fallahpour, R.A. Synthesis of 4′-substituted-2,2′:6′,2″-terpyridines. *Synthesis* **2003**, *2*, 155–184. [CrossRef]

15. Thompson, A.M.W.C. The synthesis of 2,2′:6′,2″-terpyridine ligands- versatile building blocks for supramolecular chemistry. *Coord. Chem. Rev.* **1997**, *160*, 1–52. [CrossRef]

16. Wang, J.; Hanan, G.S. A Facile Route to Sterically Hindered and Non-Hindered 4′-Aryl-2,2′:6′,2″-Terpyridines. *Synlett* **2005**, *8*, 1251–1254. [CrossRef]

17. Organic Syntheses. Available online: http://www.orgsyn.org/instructions.aspx (accessed on 9 October 2020).

18. Pinciroli, V.; Biancardi, V.; Visentin, G.; Rizzo, V. The Well-Characterized Synthetic Molecule: A Role for Quantitative ^{1}H NMR. *Org. Process. Res. Dev.* **2004**, *8*, 381–384. [CrossRef]

19. Husson, J.; Guyard, L. 4′-(5-Methylfuran-2-yl)-2,2′:6′,2″-terpyridine: A New Ligand Obtained from a Biomass-Derived Aldehyde with Potential Application in Metal-Catalyzed Reactions. *Molbank* **2018**, *2018*, M1032. [CrossRef]

20. Beley, M.; Delabouglise, D.; Houppy, G.; Husson, J.; Petit, J.-P. Preparation and properties of ruthenium (II) complexes of 2,2′:6′,2″-terpyridines substituted at the 4′-position with heterocyclic groups. *Inorg. Chim. Acta* **2005**, *358*, 3075–3083. [CrossRef]

21. Husson, J.; Dehaudt, J.; Guyard, L. Preparation of carboxylate derivatives of terpyridine via the furan pathway. *Nat. Protoc.* **2014**, *9*, 21–26. [CrossRef]

22. Et Taouil, A.; Husson, J.; Guyard, L. Synthesis and characterization of electrochromic [Ru(terpy)$_2$ selenophene]-based polymer film. *J. Electroanal. Chem.* **2014**, *728*, 81–85. [CrossRef]

23. Husson, J.; Abdeslam, E.T.; Guyard, L. A missing member in the family of chalcogenophene-substituted 2,2′:6′,2″-terpyridine: 4′-(tellurophen-2-yl)-2,2′:6′,2″-terpyridine, its Ru(II) complex and its electropolymerization as a thin film. *J. Electroanal. Chem.* **2019**, *855*, 113594. [CrossRef]

24. Zheng, C.; Pu, S.; Xu, J.; Luo, M.; Huang, D.; Shen, L. Synthesis and the effect of alkyl chain length onoptoelectronic properties of diarylethene derivatives. *Tetrahedron Lett.* **2007**, *63*, 5437–5449. [CrossRef]

25. Howbert, J.J.; Mohamadi, F.; Spees, M.M. Antitumor Compositions and methods of Treatment. U.S. Patent 5,302,724, 12 April 1994.

26. Zhang, S.; Huang, S.; Feng, C.; Cai, J.; Chen, J.; Ji, M. Novel Preparation of Tiaprofenic Acid. *J. Chem. Res.* **2013**, *37*, 406–408. [CrossRef]

MDPI

Communication

Synthesis of Polar Aromatic Substituted Terminal Alkynes from Propargyl Amine

Surya R. Banks [1], Kyung Min Yoo [1] and Mark E. Welker [2,*

[1] Department of Chemistry, Wake Forest University, Winston-Salem, NC 27101, USA; suryarbanks@gmail.com (S.R.B.); kyminyoo@gmail.com (K.M.Y.)
[2] Center for Functional Materials, Department of Chemistry, Wake Forest University, Winston-Salem, NC 27101, USA
* Correspondence: welker@wfu.edu; Tel.: +1-336-702-1953

Abstract: A series of small molecules containing polar aromatic substituents and alkynes have been synthesized. One–pot preparations of polar aromatic molecules containing an alkynyl imine and alkynyl amide are reported. A one-pot preparation of a catechol containing an alkynyl amine was also attempted but in our hands it proved much better to synthesize this target molecule via a three step synthesis which we also report here.

Keywords: catechol; alkyne; thiol-alkyne click reaction

check for updates

Citation: Banks, S.R.; Yoo, K.M.; Welker, M.E. Synthesis of Polar Aromatic Substituted Terminal Alkynes from Propargyl Amine. *Molbank* **2021**, *2021*, M1206. https://doi.org/10.3390/M1206

Academic Editors: Dimitrios Matiadis, Rodrigo Abonia and Eleftherios Halevas

Received: 20 March 2021
Accepted: 19 April 2021
Published: 25 April 2021

Publisher's Note: MDPI stays neutral with regard to jurisdictional claims in published maps and institutional affiliations.

1. Introduction

A series of small molecules containing polar aromatic substituents and propargyl amines were synthesized so that they could potentially be incorporated into hydrogel systems as an approach to developing a better hydrogen bonded and more rigid hydrogel via a thiol-alkyne click reaction [1–6]. Propargyl amines are also an important class of molecules in their own right, and are used as building blocks in heterocyclic chemistry and pharmaceutical chemistry [7,8]. Three main structural aspects of the small molecules to be synthesized were considered: (1) a polar functional group for enhanced hydrogen bonding, (2) an alkyne functional group for attachment to thiol containing hydrogels via the thiol-alkyne click reaction, and (3) ease of synthesis of the small molecule, i.e., where possible, one-pot reactions from inexpensive, commercially available starting materials.

2. Results and Discussion

To satisfy the above criteria, we initially performed reactions between substituted benzaldehydes (**1**) with propargyl amine (**2**) as shown in Scheme 1. Condensation of propargyl amine (**2**) with the aldehydes (**1**) led to the imines (**3–8**) in good yield.

3 R_1 = OH, R_2 = OH, 85%
4 R_1 = H, R_2 = CO_2H, 93%
5 R_1 = F, R_2 = OH, 90%
6 R_1 = H, R_2 = $NHC(O)CH_3$, 87%
7 R_1 = H, R_2 = CO_2CH_3, 89%
8 R_1 = H, R_2 = OH, 88%

Scheme 1. Preparation of propargyl imines from propargyl amine.

Likewise, the coupling of propargyl amine (**2**) with the benzoic acids (**9**) using EDC as a coupling agent led to the amides (**10**, **11**) also in good isolated yield (Scheme 2).

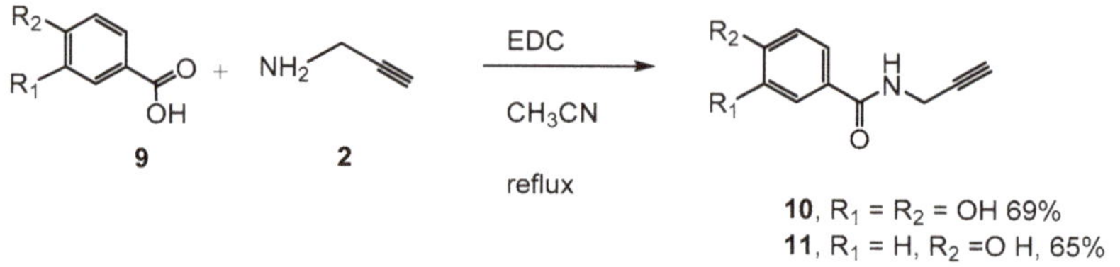

Scheme 2. Preparation of propargyl amides from propargylamine.

Preparation of the amine (**15**) proved much more complicated. There is a reported literature procedure for reductive amination of 3,4-dihydroxybenzaldehyde with propargyl amine [9], but the reported yield is low (31%). The product is reported as a red solid, which seems unlikely for a pure compound with just a benzene ring or alkyne as a chromophore and it could be that this material also contains some charge transfer complexes produced under these conditions. When we performed the literature reaction, we isolated mixtures of amine **15** and what we think may possibly be the catecholboronate dimer. Rather than spend a lot of time trying to rigorously identify this byproduct, we chose to investigate an alternate, straightforward route for the preparation of compound **15** (Scheme 3. See Supplementary Materials). Ultimately, to obtain pure amine (**15**), we found that we first had to protect the catechol (**1**, R$_1$ = R$_2$ = OH) as previously reported acetonide (**12**) [10,11], which was then subjected to reductive amination to produce (**13**) as shown in Scheme 3. Acetonide (**13**) was deprotected to yield ammonium salt (**14**) which was then deprotonated to yield the desired amine (**15**).

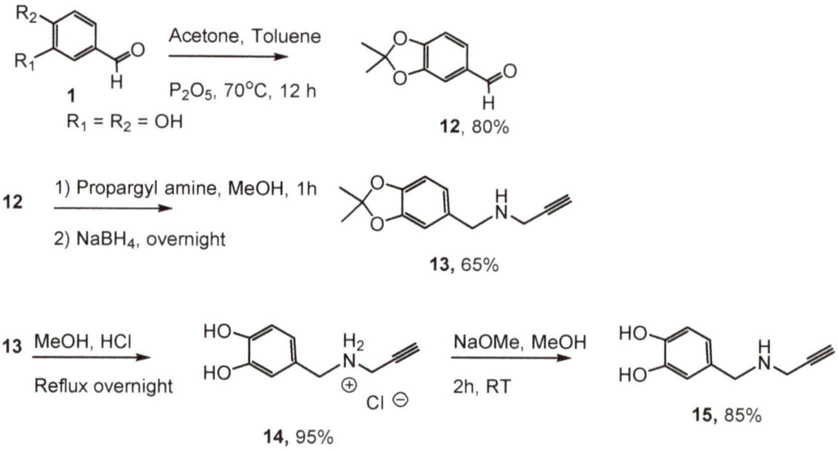

Scheme 3. Optimized preparation of amine **15**.

3. Experimental

General Methods

NMR spectra were obtained on a Bruker 400 MHz spectrometer and mass spectrometry was performed on a Thermo LTQ Orbitrap XL. All reagents and materials were obtained from the suppliers listed below. Fischer Scientific: sodium sulfate; Acros: 1,2-Dichloroethane, 1-Ethyl-3-(3-dimethylaminopropyl)carbodiimide, propargylamine; Sigma Aldrich: all aromatic aldehydes and acids; Cambridge Isotope Laboratories: Dimethyl Sulfoxide-d_6 + 0.05% v/v TMS. Compounds **3**, **10** and **15** are also described in a patent [12].

(E)-4-((Prop-2-yn-1-ylimino)methyl)benzene-1,2-diol (**3**). To a solution of 3,4-dihydroxybenzlaldehyde (0.200 g, 1.45 mmol) in 5:1 DCE:THF (6 mL), propargylamine (**2**) (111 μL, 1.74 mmol, 1.2 eq) was added dropwise. Fifteen minutes into the reaction, a grey solid started to precipitate. The reaction mixture was stirred for 2 h at room temperature under nitrogen. The resulting precipitate was filtered under vacuum, washed with 5:1 DCE:THF solution (3 × 5 mL), and dried under high vacuum. Compound **3** was isolated as a tan solid (0.218 g, 1.24 mmol, 85%). ^{1}H-NMR (400 MHz, DMSO-d_6) δ: 9.14 (br s, 2H), 8.29 (s, 1H), 7.21 (s, 1H), 7.00 (d, J = 8.0 Hz, 1H), 6.78 (d, J = 8.0 Hz, 1H), 4.38 (dd, J = 2.5, 1.7 Hz, 2H), 3.37 (t, J = 2.5 Hz, 1H). ^{13}C-NMR (101 MHz, DMSO-d_6) δ: 162.17, 149.01, 146.02, 127.87, 121.82, 115.81, 114.20, 80.97, 77.36, 47.08. HRMS (EI) for $C_{10}H_9NO_2$ 176.0712 [M + H]$^+$, found 176.0713.

(E)-4-((prop-2-yn-1-ylimino)methyl)benzoic acid (**4**). To a solution of 4-carboxybenzadehyde (0.200 g, 1.33 mmol) in 1,2-dichloroethane (10 mL), propargylamine (103 μL, 1.60 mmol, 1.2 eq) was added dropwise and the reaction was magnetically stirred under nitrogen overnight at room temperature. The resulting precipitate was filtered, washed with DCE (3 × 5 mL), and dried under high vacuum. Compound **4** was isolated as a light tan solid (0.232 g, 1.23 mmol, 93%). ^{1}H-NMR (400 Hz, DMSO-d_6) δ: 8.62 (t, J = 1.9 Hz, 1H), 8.01 (m, 2H), 7.88 (m, 2H), 4.55 (m, 2H), 3.48 (t, J = 2.5 Hz, 1H). ^{13}C-NMR (101 MHz, DMSO-d_6) δ: 167.52, 161.87, 139.43, 133.93, 130.10, 128.50, 80.15, 78.19, 47.36. HRMS (EI) for $C_{11}H_9NO_2$ 188.0712 [M − H]$^+$, found 188.0705.

(E)-2-fluoro-4-((prop-2-yn-1-ylimino)methyl)phenol (**5**). To a solution of 3-fluoro-4-hydroxybenzadehyde (0.200 g, 1.43 mmol) in a 5:1 DCE:THF (6 mL), propargylamine (110 μL, 1.46 mmol, 1.2 eq) was added dropwise and the reaction was magnetically stirred under nitrogen overnight at room temperature. The reaction solvent was then dried with sodium sulfate, filtered, and evaporated in vacuo and the obtained product was dried under high vacuum. Compound **5** was isolated as a solid (0.228 g, 1.28 mmol, 90%). ^{1}H-NMR (400 MHz, DMSO-d_6) δ: 10.39 (br s, 1H), 8.39 (s, 1H), 7.53 (d, J = 12.0, 1.9 Hz, 1H), 7.41 (d, J = 8.4, 1.9 Hz, 1H), 7.02 (t, J= 8.6 Hz, 1H), 4.43 (s, 2H), 3.41 (t, J = 2.4 Hz, 1H).^{13}C-NMR (101 MHz, DMSO-d_6) δ: 161.09, 151.6 (d, J = 243 Hz), 150.39, 148.31, 128.03, 125.93, 118.12, 115.23 (d, J = 19 Hz), 80.57, 77.73, 46.96. HRMS (EI) for $C_{10}H_9FNO$ 178.0668 [M − H]$^+$, found 178.0670.

N-(4-((prop-2-yn-1-ylimino)methyl)phenyl)acetamide (**6**). To a solution of 4-acetamidobenzaldehyde (0.200 g, 1.23 mmol) in 5:1 DCE:THF (6 mL), propargylamine (94 μL, 1.47 mmol, 1.2 eq) was added dropwise and the reaction was magnetically stirred under nitrogen overnight at room temperature. The reaction solvent was then dried with sodium sulfate, filtered then evaporated in vacuo and the obtained product was dried under high vacuum. Compound **6** was isolated as a light yellow solid (0.215 g, 1.07 mmol, 87%). ^{1}H-NMR (400 MHz, DMSO-d_6) δ: 10.16 (s, 1H), 8.45 (t, J = 1.8 Hz, 1H), 7.68 (m, 4H), 4.45 (br s, 2H), 3.41 (t, J = 2.5 Hz, 1H), 2.08 (s, 3H). ^{13}C-NMR (101 MHz, DMSO-d_6) δ: 169.11, 161.80, 142.22, 130.81, 129.23, 119.14, 80.62, 77.78, 47.15, 24.58. HRMS (EI) for $C_{12}H_{13}N_2O$ 201.1028 [M − H]$^+$, found 201.1030.

Methyl (E)-4-((prop-2-yn-1-ylimino)methyl)benzoate (**7**). To a solution of methyl-4-formyl-benzoate (0.200 g, 1.22 mmol) in 1,2-dichloroethane (6 mL), propargylamine (94 μL, 1.47 mmol, 1.2 eq) was added dropwise and the reaction was magnetically stirred under nitrogen overnight at room temperature. The reaction solvent was then dried with sodium sulfate, filtered, and evaporated in vacuo and the obtained product was dried under high vacuum. Compound **7** was isolated as an off-white solid (0.219 g, 1.08 mmol, 89%). ^{1}H-NMR (400 MHz, DMSO-d_6) δ: 8.63 (s, 1H), 8.04 (d, J = 8.2 Hz, 2H), 7.92 (d, J = 8.3 Hz, 2H), 4.56 (t, J = 2.2 Hz, 2H), 3.88 (s, 3H), 3.49 (t, J = 2.5 Hz, 1H). ^{13}C-NMR (101 MHz, DMSO-d_6) δ: 166.26, 161.71, 140.04, 131.91, 130.01, 128.71, 80.08, 78.26, 52.78, 47.39. HRMS (EI) for $C_{12}H_{11}NO_2$ 202.0863 [M − H]$^+$, found 202.0870.

(E)-4-((prop-2-yn-1-ylimino)methyl)phenol (**8**). To a solution of 4-hydroxybenzadehyde (0.200 g, 1.64 mmol) in 1,2-dichloroethane (6 mL), propargylamine (126 µL, 1.97 mmol, 1.2 eq) was added dropwise and the reaction was magnetically stirred under nitrogen overnight at room temperature. The reaction solvent was then dried with sodium sulfate, filtered, and evaporated in vacuo and the obtained product was dried under high vacuum. Compound **8** was isolated as a light red solid (0.232 g, 1.45 mmol, 88%). ^{1}H-NMR (400 MHz, DMSO-d_6) δ: 9.85 (s, 1H), 8.39 (t, J = 1.7 Hz, 1H), 7.60 (d, J = 8.0 Hz, 2H), 6.83 (d, J = 8.0 Hz, 2H), 4.40 (t, J = 2.5, 1.7 Hz, 2H), 3.38 (t, J = 2.5 Hz, 1H). ^{13}C-NMR (126 MHz, DMSO) δ: 161.90, 160.52, 130.27, 127.48, 115.96, 80.87, 77.50, 47.10. HRMS (EI) for $C_{10}H_9NO$ 160.0672 $[M - H]^+$, found 159.9674.

3,4-Dihydroxy-N-(prop-2-yn-1-yl)benzamide (**10**). To a stirred solution of 3,4-dihydroxybenzoic acid (**8**, R_1 = R_2 = OH) (0.200 g, 1.30 mmol, 1.0 eq) in acetonitrile (10 mL), propargylamine (**2**) (165 µL, 2.55 mmol, 2.0 eq) was added dropwise, resulting in a white precipitate. Subsequently, 1-Ethyl-3-(3-dimethylaminopropyl)carbodiimide (0.240 g, 1.55 mmol, 1.2 eq) was added slowly and the reaction mixture was refluxed overnight with magnetic stirring. Following the removal of the reaction solvent in vacuo, the product was extracted with ethyl acetate (3 × 20 mL) from water (10 mL). The combined organic layers were dried over anhydrous Na_2SO_4 and the solvent was removed in vacuo. The crude product was purified by column chromatography on silica gel (3:4 EtOAc/Hex) to obtain compound **10** as a tan-colored solid (0.209 g, 1.09 mmol, 69%): ^{1}H-NMR (400 MHz, DMSO-d_6) δ: 9.31 (br s, 2H), 8.56 (s, 1H), 7.29 (s, 1H), 7.20 (d, J = 8.3 Hz, 1H), 6.76 (d, J = 8.3 Hz, 1H), 3.99 (m, 2H), 3.06 (t, J = 2.5 Hz, 1H). ^{13}C-NMR (400 MHz, DMSO-d_6) δ: 166.25, 149.02, 145.32, 125.52, 119.51, 115.60, 115.33, 82.20, 72.94, 28.82. HRMS (ESI-TOF) for $C_{10}H_9NO_3$ 192.0661 $[M + H]^+$, found 192.0663.

4-hydroxy-N-(prop-2-yn-1-yl)benzamide (**11**). This compound has been prepared previously by an alternate procedure [13]. To a stirred solution of 4-hydroxybenzoic acid (**8**, R_1 = OH, R_2 = H) (0.200 g, 1.45 mmol, 1.0 eq) in acetonitrile (10 mL), propargylamine (111 µL, 1.74 mmol, 1.2 eq) was added dropwise, resulting in a white precipitate. Subsequently, 1-Ethyl-3-(3-dimethylaminopropyl)carbodiimide (0.270 g, 1.74 mmol, 1.2 eq) was added slowly and the reaction mixture was refluxed overnight with magnetic stirring. Following the removal of the reaction solvent in vacuo, the product was extracted with ethyl acetate (3 × 20 mL) from water (10 mL). The combined organic layers were dried over anhydrous Na_2SO_4 and the solvent was removed in vacuo. The crude product was purified by column chromatography on silica gel (1:1 EtOAc/Hex) to obtain compound **11** as a tan-colored solid (0.185 g, 1.06 mmol, 73%) ^{1}H-NMR (400 MHz, DMSO-d_6) δ: 10.00 (s, 1H), 8.64 (t, J = 5.5 Hz, 1H), 7.72 (d, J = 8.0 Hz, 2H), 6.80 (d, J = 8.0, 2H), 4.01 (dd, J = 5.6, 2.5 Hz, 2H), 3.08 (t, J = 2.5 Hz, 1H). ^{13}C-NMR (101 MHz, DMSO-d_6) δ: 166.05, 160.75, 129.67, 124.96, 115.29, 82.13, 73.05, 28.80.

2,2-Dimethylbenzo[d][1,3]dioxole-5-carbaldehyde (**12**). A mixture of 3,4-dihydroxybenzaldehyde (**1**) (0.276 g, 2 mmol) and P_2O_5 (0.141 g, 1 mmol) was stirred in toluene (dry) (100 mL). Acetone (0.74 mL, 10 mmol) was added and the mixture stirred at 75 °C for 3 h. Four portions of P_2O_5 (4 × 0.100 g) were added every 30 min during heating. The reaction was quenched with 25% NaOH (aq) (25 mL) and the toluene solvent removed under vacuum after separation. The crude solid obtained was purified by column chromatography (DCM:Hexane, 2:1) to obtain a light brown solid (**12**) (0.300 g, 1.6 mmol, 80%) identical by NMR comparison to previously reported material [10,11]: ^{1}H-NMR (400 MHz, DMSO-d_6) δ 9.79 (s, 1H), 7.51 (dd, J = 8.0, 1.6 Hz, 1H), 7.26 (d, J = 1.6 Hz, 1H), 7.06 (d, J = 8.0 Hz, 1H), 1.69 (s, 6H).

N-((2,2-Dimethylbenzo[d][1,3]dioxol-5-yl)methyl)prop-2-yn-1-amine (**13**). Compound **12** (0.250 g, 1.4 mmol) was dissolved in methanol (dry) (25 mL) and propargylamine (**3**) (0.13 mL, 2.0 mmol) was added dropwise. The solution was allowed to stir for 1 h and sodium borohydride (3 × 0.100 g, 8.0 mmol) was added in portions over 30 min. The reaction was stirred overnight under nitrogen atmosphere and quenched with brine (15 mL). The aqueous was extracted using ethyl acetate (25 mL) and solvent dried (Na$_2$SO$_4$), and evaporated in vacuo to obtain a viscous liquid (**13**) (0.197 g, 0.91 mmol, 65%): ^{1}H-NMR (400 MHz, DMSO-d_6) δ 6.78 (dd, J = 1.4, 0.7 Hz, 1H), 6.72 (dd, J = 2.9, 1.0 Hz, 2H), 3.62 (s, 2H), 3.24 (d, J = 2.4 Hz, 2H), 3.06 (t, J = 2.4 Hz, 1H), 1.62 (s, 6H). ^{13}C-NMR (101 MHz, DMSO-d_6) δ: δ 147.31, 146.09, 133.86, 121.01, 118.01, 108.70, 108.02, 83.33, 74.13, 51.57, 36.90, 26.00. HRMS (ESI-TOF) for C$_{13}$H$_{16}$NO$_2$ 218.1181 [M + H]$^+$, found 218.1171.

N-(3,4-Dihydroxybenzyl)prop-2-yn-1-ammonium chloride (**14**). Compound **13** (0.180 g, 0.82 mmol) was dissolved in anhydrous methanol (20 mL) and purged with dry HCl for 5 min. The solution was refluxed overnight. The solvents were evaporated and the solid triturated with diethyl ether (3 × 5 mL) to obtain blue-black solid (**14**) (0.165 g, 0.77 mmol, 94%): ^{1}H-NMR (400 MHz, DMSO-d_6) δ 9.67 (s, 2H), 7.08–6.66 (m, 3H), 3.96 (brs, 2H), 3.78 (brs, 2H), 3.73 (brs, 1H), 3.17 (brs, 1H). ^{13}C-NMR (101 MHz, DMSO-d_6) δ: 146.68, 145.76, 122.25, 121.87, 118.11, 116.12, 80.06, 75.46, 49.32, 35.20. HRMS (ESI-TOF) for C$_{10}$H$_{12}$NO$_2$ 178.0868 [M + H]$^+$, found 178.0862.

4-((Prop-2-yn-1-ylamino)methyl)benzene-1,2-diol (**15**). Compound **14** (0.150 g, 0.7 mmol) was dissolved in methanol (anhydrous) (20 mL) and sodium methoxide (0.040 g, 0.74 mmol) was added. The mixture was stirred for 2 h at room temperature and quenched with DI water 1 mL). The mixture was then filtered through celite and the solvent was removed in vacuo to yield a brown solid (**15**) (0.105 g, 0.593 mmol, 85%): ^{1}H-NMR (400 MHz, DMSO-d_6) δ 6.71 (d, J = 2.1 Hz, 1H), 6.65 (d, J = 7.9 Hz, 1H), 6.52 (dd, J = 8.0, 2.1 Hz, 1H), 3.54 (s, 2H), 3.22 (d, J = 2.5 Hz, 2H), 3.05 (t, J = 2.4 Hz, 1H). ^{13}C-NMR (101 MHz, DMSO-d_6) δ: 145.03, 144.06, 130.74, 118.84, 115.78, 115.32, 82.97, 73.55, 50.93, 36.35. HRMS (ESI-TOF) for C$_{10}$H$_{12}$NO$_2$ 178.0868 [M + H]$^+$, found 178.0868.

4. Conclusions

We successfully prepared a number of new polar aromatic substituted terminal alkynes from propargyl amine and we hope scientists working with thiolated hydrogels will incorporate them into their hydrogels and that they will also be used in pharmaceutical chemistry, with the anticipation that those modified molecules will have interesting new properties.

Supplementary Materials: The following data are available online: ^{1}H and ^{13}C-NMR spectra for compounds **3–15**.

Author Contributions: K.M.Y. prepared compounds **3–11**; S.R.B. prepared compounds **12–15**; K.M.Y., S.R.B. and M.E.W. analyzed the spectral data and wrote the manuscript. All authors have read and agreed to the published version of the manuscript.

Funding: This research received no external funding.

Data Availability Statement: The data presented in this study are available upon request from the corresponding author.

Acknowledgments: The authors thank Wake Forest University for internal pilot funding of this work.

Conflicts of Interest: The authors declare no conflict of interest.

References

1. Heida, T.; Otto, O.; Biedenweg, D.; Hauck, N.; Thiele, J. Microfluidic Fabrication of Click Chemistry-Mediated Hyaluronic Acid Microgels: A Bottom-Up Material Guide to Tailor a Microgel's Physicochemical and Mechanical Properties. *Polymers* **2020**, *12*, 1760. [CrossRef] [PubMed]
2. Xu, Z.; Bratlie, K.M. Click Chemistry and Material Selection for in Situ Fabrication of Hydrogels in Tissue Engineering Applications. *ACS Biomater. Sci. Eng.* **2018**, *4*, 2276–2291. [CrossRef] [PubMed]
3. Gopinathan, J.; Noh, I. Click Chemistry-Based Injectable Hydrogels and Bioprinting Inks for Tissue Engineering Applications. *Tissue Eng. Regen. Med.* **2018**, *15*, 531–546. [CrossRef] [PubMed]
4. Meng, X.; Edgar, K.J. "Click" reactions in polysaccharide modification. *Prog. Polym. Sci.* **2016**, *53*, 52–85. [CrossRef]
5. Sarkar, B.; Jayaraman, N. Glycoconjugations of Biomolecules by Chemical Methods. *Front. Chem.* **2020**, *8*, 888. [CrossRef] [PubMed]

6. Ma, H.; Caldwell, A.S.; Azagarsamy, M.A.; Rodriguez, A.G.; Anseth, K.S. Bioorthogonal click chemistries enable simultaneous spatial patterning of multiple proteins to probe synergistic protein effects on fibroblast function. *Biomaterials* **2020**, *255*, 120205. [CrossRef] [PubMed]
7. Zorba, L.P.; Vougioukalakis, G.C. The Ketone-Amine-Alkyne (KA2) coupling reaction: Transition metal-catalyzed synthesis of quaternary propargylamines. *Coord. Chem. Rev.* **2021**, *429*, 213603. [CrossRef]
8. Ghosh, S.; Biswas, K. Metal-free multicomponent approach for the synthesis of propargylamine: A review. *RSC Adv.* **2021**, *11*, 2047–2065. [CrossRef]
9. Finbloom, J.A.; Han, K.; Slack, C.C.; Furst, A.L.; Francis, M.B. Cucurbit[6]uril-Promoted Click Chemistry for Protein Modification. *J. Am. Chem. Soc.* **2017**, *139*, 9691–9697. [CrossRef] [PubMed]
10. Tran, A.T.; West, N.P.; Britton, W.J.; Payne, R.J. Elucidation of Mycobacterium tuberculosis Type II Dehydroquinase Inhibitors using a Fragment Elaboration Strategy. *ChemMedChem* **2012**, *7*, 1031–1043. [CrossRef] [PubMed]
11. Kim, J.-K.; Noh, J.H.; Lee, S.; Choi, J.S.; Suh, H.; Chung, H.Y.; Song, Y.; Choi, W.C. The First Total Synthesis of 2,3,6-Tribromo-4,5-dihydroxybenzyl Methyl Ether (TDB) and Its Antioxidant Activity. *Bull. Korean Chem. Soc.* **2002**, *23*, 661–662. [CrossRef]
12. Welker, M.E.; Skardal, A.; Weissenfluh, A.; Banks, S. Hydrogen-Bonding Compounds, Compositions Comprising the Same, and Methods of Preparing and Using the Same. U.S. Patent No. US20190345096A1, 14 November 2019.
13. Gopin, A.; Ebner, S.; Attali, B.; Shabat, D. Enzymatic Activation of Second-Generation Dendritic Prodrugs: Conjugation of Self-Immolative Dendrimers with Poly(ethylene glycol) via Click Chemistry. *Bioconjug. Chem.* **2006**, *17*, 1432–1440. [CrossRef] [PubMed]

Communication

Synthesis and Structural Characterization of (*E*)-4-[(2-Hydroxy-3-methoxybenzylidene)amino]butanoic Acid and Its Novel Cu(II) Complex

Eleftherios Halevas [1,*], Antonios Hatzidimitriou [2], Barbara Mavroidi [1], Marina Sagnou [1], Maria Pelecanou [1] and Dimitris Matiadis [1,*]

[1] National Centre for Scientific Research "Demokritos", Institute of Biosciences & Applications, 15310 Athens, Greece; bmavroidi@bio.demokritos.gr (B.M.); sagnou@bio.demokritos.gr (M.S.); pelmar@bio.demokritos.gr (M.P.)

[2] Laboratory of Inorganic Chemistry, Department of Chemistry, Aristotle University of Thessaloniki, 54124 Thessaloniki, Greece; hatzidim@chem.auth.gr

* Correspondence: lefterishalevas@gmail.com (E.H.); dmatiadis@gmail.com or matiadis@bio.demokritos.gr (D.M.); Tel.: +30-210-6503558 (D.M.)

Abstract: A novel Cu(II) complex based on the Schiff base obtained by the condensation of *ortho*-vanillin with *gamma*-aminobutyric acid was synthesized. The compounds are physico-chemically characterized by elemental analysis, HR-ESI-MS, FT-IR, and UV-Vis. The complex and the Schiff base ligand are further structurally identified by single crystal X-ray diffraction and ^{1}H and ^{13}C-NMR, respectively. The results suggest that the Schiff base are synthesized in excellent yield under mild reaction conditions in the presence of glacial acetic acid and the crystal structure of its Cu(II) complex reflects an one-dimensional polymeric compound. The molecular structure of the complex consists of a Cu(II) ion bound to two singly deprotonated Schiff base bridging ligands that form a CuN_2O_4 chelation environment, and a coordination sphere with a disordered octahedral geometry.

Keywords: imine; Schiff base; X-ray crystallographic analysis; Cu(II) complex; *gamma*-amino acid

Citation: Halevas, E.; Hatzidimitriou, A.; Mavroidi, B.; Sagnou, M.; Pelecanou, M.; Matiadis, D. Synthesis and Structural Characterization of (*E*)-4-[(2-Hydroxy-3-methoxybenzylidene)amino]butanoic Acid and Its Novel Cu(II) Complex. *Molbank* **2021**, *2021*, M1179. https://doi.org/10.3390/M1179

Academic Editor: Kristof Van Hecke

Received: 23 December 2020
Accepted: 2 January 2021
Published: 6 January 2021

Publisher's Note: MDPI stays neutral with regard to jurisdictional claims in published maps and institutional affiliations.

1. Introduction

Schiff bases have attracted intensive scientific interest mainly because of their ease of preparation, diverse structural and physico-chemical characteristics, metal binding affinity, pharmacological and physiological properties. They have been extensively used as ligands in the coordination chemistry of main group and transition metal ions. These compounds and their metal complexes have been proved very effective as catalysts in several biological systems, dyes, polymers, and as bioactive agents in the pharmaceutical and medicinal fields presenting exceptional antibacterial, antioxidant, antifungal, anticancer, antidiabetic, and diuretic properties [1–5].

A well-established sub-category of Schiff bases is the amino acid-based Schiff bases. The incorporation of amino acids in the Schiff base structure enables the design of ligands with enhanced chirality and multidentate functionality [6]. Several literature reports confirm the involvement of amino acid Schiff base complexes in a variety of chemical and biological processes such as the catalysis of transamination, carboxylation and racemization reactions [7], the oxidation of sulfides and olefins, in polymerization processes, and the decomposition of H_2O_2 [8]. Moreover, it has been proved that amino acid Schiff base complexes can be used as radiotracers in nuclear medicine [9], and as anticancer and antibacterial agents [10,11].

Until today, a large number of crystallographically characterized metal complexes of *alpha*- and *beta*-amino acid Schiff bases have been reported in the literature [12,13]. However, only a few metal complexes of Schiff bases derived from *gamma*-amino acids have been structurally characterized via X-ray crystallography [14]. Herein, we report the synthesis of a novel Cu(II) complex based on a new Schiff base obtained by the condensation of *ortho*-vanillin with *gamma*-aminobutyric acid. The compounds have been physico-chemically characterized by elemental analysis, HR-ESI-MS, FT-IR, and UV-Visible spectrophotometry. The complex and the Schiff base ligand have been further structurally identified by single crystal X-ray and ^{1}H and ^{13}C-NMR, respectively.

2. Results and Discussion

2.1. Synthesis

The Schiff base **3** was synthesized in one step by mixing equimolar quantities of *ortho*-vanillin **1** and *gamma*-aminobutyric acid **2** in the minimum amount of refluxing MeOH (Scheme 1). After filtration and washing, product **3** was obtained in very good yield (76%). In order to further improve the yield of the reaction, the application of (a) glacial acetic acid as a mild acidic agent appropriate for such couplings [15,16], (b) sulfuric acid as a dehydrating acid [17], and (c) piperidine as one the most commonly used bases for Schiff base formation of vanillin derivatives [18] was explored (Table 1). Piperidine seemed to partially "quench" the coupling, leading to lower yields, even after forcing the product precipitation. The presence of glacial acetic acid increased the yield of the reaction to almost quantitative (86%), whereas using concentrated sulfuric acid was proved to be unsuccessful.

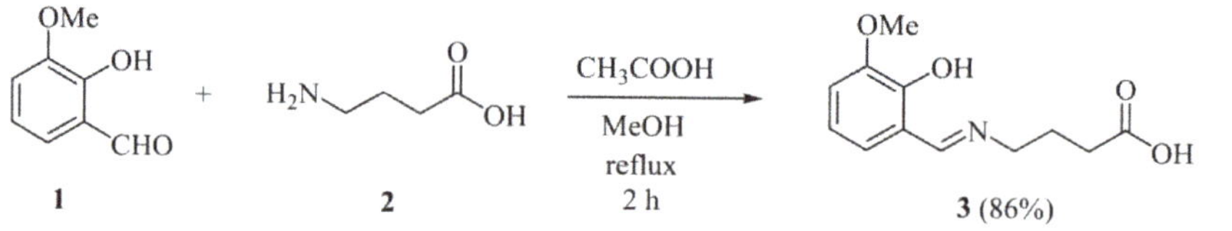

Scheme 1. Synthesis of (*E*)-4-((2-hydroxy-3-methoxybenzylidene)amino)butanoic acid (**3**).

Table 1. Optimization of the reaction conditions regarding the additive [a].

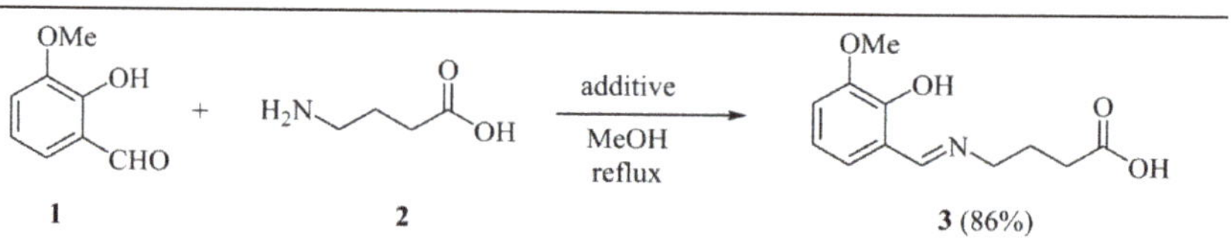

Entry	Additive	Yield (%) [b]
1	none	76
2	CH$_3$COOH	86
3	H$_2$SO$_4$ [c]	trace amount
4	piperidine	58 [c]

[a] The reactions were carried out using 1.0 mmol of each starting compound and 2 drops of each additive (0.1 mL), under reflux for 2 h. [b] Isolated yield after concentration of the solvent to 2/3 of its volume. [c] Isolated yield, after further concentration of the solvent to 1/3 of its volume and storage at −18 °C for 48 h.

The structure and purity of compound **3** was confirmed by ^{1}H and ^{13}C-NMR (Supplementary Information, Figures S1–S6). As expected, the methylene protons of the amino acid unit appeared in the aliphatic region of the ^{1}H-NMR spectrum. H-4 protons (see Figure 1A numbering) were shifted downfield at 3.60 ppm because of the nitrogen atom. The aromatic protons appeared according to the standard shift and splitting pattern of 1,2,3-trisubstituted benzene rings at 6.79, 7.00, and 7.02 ppm. The ^{13}C-NMR spectrum is in good agreement with the structure and reported data [19,20]. The low-field signals of azomethine, C-7 and carboxylic carbons were assigned by 2D-NMR experiments (Supplementary Information, Figures S3–S6) at 166.2, 151.9, and 174.0 ppm, respectively.

NOESY experiment demonstrated that **3** has the same configuration with the coordinated ligand, since a signal corresponding to azomethine proton (8.53 ppm) and H-9 of the ring (7.02 ppm) proximity was detected, thus proving the expected *E*-configuration.

Under standard NMR experiment conditions (0.05 M in DMSO-d_6) we were not able to distinguish the phenolic OH proton from the carboxylic OH proton, due to broadening of these peaks, leading to the absence of 2D signals. Even by modifying the conditions (concentration, temperature, and/or solvent) we did not observe the desired peak sharpening or appearance of any additional clarifying 2D signals. Since the structural elucidation by NMR of similar Schiff bases systems incorporating an *ortho*-hydroxy phenyl unit have been reported in the literature [19,20], the broad peak at 12.2 ppm was assigned to the carboxylic proton, whereas the lower-field broad peak at 13.7 ppm was assigned to the deshielded phenolic proton.

For the synthesis of the heteroleptic Cu(II)-Schiff base complex (**4**) (Scheme 2) a thorough exploration of the experimental conditions regarding solvent system, pH, temperature, metal:ligand stoichiometry, and crystallization process was performed. Subsequently, complex **4** was synthesized from the reaction of Cu(NO$_3$)$_2$·3H$_2$O with the Schiff base in MeOH at 60 °C under reflux conditions, in the presence of sodium hydroxide (NaOH). The overall stoichiometric reaction leading to complex **4** is shown schematically below:

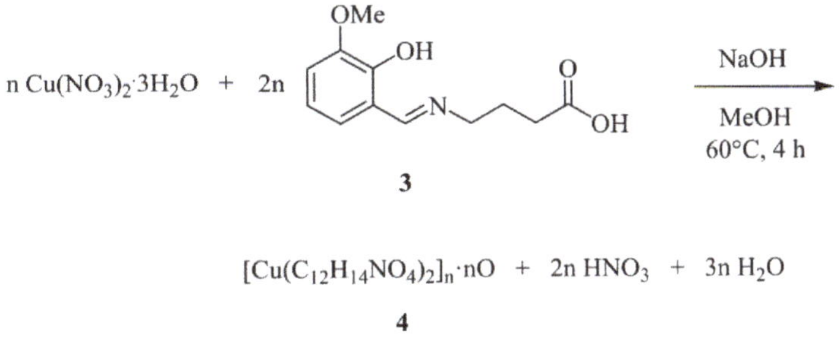

$$[Cu(C_{12}H_{14}NO_4)_2]_n \cdot nO \ + \ 2n\ HNO_3 \ + \ 3n\ H_2O$$

4

Scheme 2. Stoichiometric reaction of the heteroleptic Cu(II)-Schiff base complex (**4**).

The reaction mixture was left to evaporate slowly at room temperature. Dark green crystalline material emerged in the reaction described above, the analytical composition of which was consistent with the formulation in complex **4**. Positive identification of the crystalline product was achieved by elemental analysis, FT-IR, HR-ESI-MS, and X-ray crystallographic analysis of isolated single crystals from complex **4**.

Both compounds are stable in air for fairly long periods of time. They readily dissolve in H$_2$O, methanol (MeOH), dimethylacetamide (DMA), dimethyl sulfoxide (DMSO), and dimethylformamide (DMF) and are insoluble in acetone, acetonitrile, and dichloromethane at room temperature.

2.2. Description of X-ray Crystallographic Structure

The X-ray crystal structure of **4** reveals a discrete solid-state lattice. The molecular structure of **2** is given in Figure 1A; selected bond distances and angles are listed in Table 2. Complex **4** crystallizes in the monoclinic space group $C2/c$. The crystal structure reflects an one-dimensional polymeric compound forming infinite chains along the c crystallographic axis. The unit cell contains four mononuclear [Cu(C$_{12}$H$_{14}$NO$_4$)$_2$] monomeric complex units and half of a badly disordered lattice MeOH molecule. The molecular structure of the monomer **4** consists of a Cu(II) ion bound to two singly deprotonated Schiff base bridging ligands which coordinate through their deprotonated phenolato oxygen atom, the imino nitrogen atom, and the double bonded oxygen atoms of the protonated carboxylic acid moieties from two neighboring molecules, thereby giving rise to a Cu^{II}N$_2$O$_4$ chelation environment, and a coordination sphere reflecting a disordered octahedral geometry

(Figure 1A). The Cu-N bond lengths are 2.005(2) Å whereas the Cu-O bond lengths are in the range between 1.9272(16) and 2.6167(19) Å. These values are very similar to the related bond distances reported in the literature [21]. Intermolecular hydrogen-bonding interactions arise between the deprotonated phenolic oxygen atoms and the protonated oxygen atoms of the carboxylic groups from the neighboring monomeric complex units enforcing the polymeric chains formation and resulting in the final 1D crystal lattice (Figure 1B, Supplementary Information Table S1).

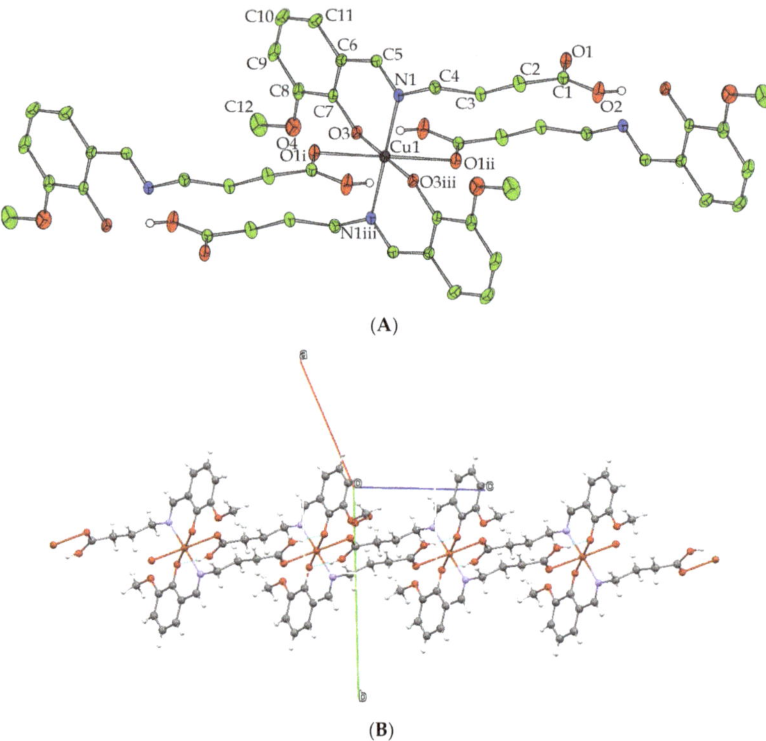

(**A**)

(**B**)

Figure 1. (**A**) The fully coordinated monomeric complex unit of **4**. Carbon group hydrogen atoms and lattice solvent molecule have been omitted. (**B**) Part of the polymeric chain of **4** and hydrogen bonding interactions (light blue dotted lines).

Table 2. Bond lengths [Å] and angles [°] for **4**.

Bond Lengths (Å)			
Cu(1)—O(1) [i]	2.6167 (19)	O(4)—C(12)	1.429 (3)
Cu(1)—O(1) [ii]	2.6167 (19)	O(5)—O(5) [iv]	1.449 (5)
Cu(1)—N(1) [iii]	2.005 (2)	C(1)—C(2)	1.505 (4)
Cu(1)—O3 [iii]	1.9272 (16)	C(2)—C(3)	1.514 (3)
Cu(1)—N(1)	2.005 (2)	C(3)—C(4)	1.510 (3)
Cu(1)—O(3)	1.9272 (16)	C(5)—C(6)	1.443 (4)
N(1)—C(4)	1.476 (3)	C(6)—C(7)	1.392 (4)
N(1)—C(5)	1.282 (3)	C(6)—C(11)	1.406 (4)
O(1)—C(1)	1.203 (3)	C(7)—C(8)	1.421 (3)
O(2)—C(1)	1.308 (3)	C(8)—C(9)	1.389 (5)
O(3)—C(7)	1.325 (3)	C(9)—C(10)	1.373 (5)
O(4)—C(8)	1.363 (4)	C(10)—C(11)	1.354 (5)

Table 2. *Cont.*

Angles (°)			
O(1) [i]—Cu(1)—O(1) [ii]	180	C(8)—O(4)—C(12)	117.5 (3)
O(1) [i]—Cu(1)—N(1) [iii]	93.65 (7)	O(2)—C(1)—O(1)	122.9 (2)
O(1) [ii]—Cu(1)—N(1) [iii]	86.35 (7)	O(2)—C(1)—C(2)	114.0 (2)
O(1) [i]—Cu(1)—O(3) [iii]	83.34 (7)	O(1)—C(1)—C(2)	123.1 (2)
O(1) [ii]—Cu(1)—O(3) [iii]	96.66 (7)	C(1)—C(2)—C(3)	112.2 (2)
N(1) [iii]—Cu(1)—O(3) [iii]	90.23 (8)	C(2)—C(3)—C(4)	112.8 (2)
O(1) [i]—Cu(1)—N(1)	86.35 (7)	C(3)—C(4)—N(1)	110.5 (2)
O(1) [ii]—Cu(1)—N(1)	93.65 (7)	N(1)—C(5)—C(6)	126.7 (2)
N(1) [iii]—Cu(1)—N(1)	180	C(5)—C(6)—C(7)	121.9 (2)
O(3) [iii]—Cu(1)—N(1)	89.77 (8)	C(5)—C(6)—C(11)	117.6 (3)
O(1) [i]—Cu(1)—O(3)	96.66 (7)	C(7)—C(6)—C(11)	120.5 (3)
O(1) [ii]—Cu(1)—O(3)	83.34 (7)	C(6)—C(7)—O(3)	123.9 (2)
N(1) [iii]—Cu(1)—O(3)	89.77 (8)	C(6)—C(7)—C(8)	118.2 (3)
O(3) [iii]—Cu(1)—O(3)	180	O(3)—C(7)—C(8)	117.8 (3)
N(1)—Cu(1)—O(3)	90.23 (8)	C(7)—C(8)—O(4)	114.6 (3)
Cu(1)—N(1)—C(4)	120.85 (16)	C(7)—C(8)—C(9)	119.2 (3)
Cu(1)—N(1)—C(5)	122.71 (18)	O(4)—C(8)—C(9)	126.2 (3)
C(4)—N(1)—C(5)	116.4 (2)	C(8)—C(9)—C(10)	121.5 (3)
Cu(1) [v]—O(1)—C(1)	124.19 (17)	C(9)—C(10)—C(11)	120.1 (3)
Cu(1)—O(3)—C(7)	123.43 (16)	C(6)—C(11)—C(10)	120.5 (3)
Cu(1)—O(1) [i]	2.6167 (19)	O(4)—C(12)	1.429 (3)
Cu(1)—O(1) [ii]	2.6167 (19)	O(5)—O(5) [iv]	1.449 (5)

Symmetry codes: [i] x, y, z − 1; [ii] −x + 3/2, −y + 3/2, −z + 3; [iii] −x + 3/2, −y + 3/2, −z + 2; [iv] −x + 1, y, −z + 1/2; [v] x, y, z + 1.

2.3. FT-IR Spectroscopy

The FT-IR spectrum of the Schiff base **3** (Supplementary Information, Figure S7) shows a weak band at 3419 cm^{-1}, characteristic of the ν(OH) vibrations [22], which disappears in the spectrum of complex **4** (Supplementary Information, Figure S7), indicating deprotonation of the OH group upon binding with the Cu(II) ion. The ν(C=N) vibrations are observed at 1643 cm^{-1} for Schiff base **3** and at 1598 cm^{-1} for complex **4**, respectively. This lowering of resonance frequency by 45 cm^{-1} for the C=N vibration clearly reveals the coordination of the Cu(II) ion with the imine nitrogen. Moreover, both spectra show broad absorption bands at about $3068–2758 \text{ cm}^{-1}$ and $3089–2818 \text{ cm}^{-1}$, respectively, which are assigned to the ν(C–H) stretching vibrations of the aromatic moieties. The absorption bands located at 526 cm^{-1} and 465 cm^{-1} in the spectrum of complex **4** can be assigned to the ν(Cu-O) and ν(Cu-N) vibrations, respectively.

2.4. UV-Vis Spectroscopy

The UV–Vis spectra of the Schiff base **3** and complex **4** were recorded in MeOH at a concentration of 10^{-5} M (Figure 2). The electronic absorption spectrum of the Schiff base shows an absorption band at 293 nm and 417 nm that can be assigned to the $\pi \to \pi^*$ transitions in the aromatic ring and $n \to \pi^*$ transitions of the imine moiety, respectively. In the spectrum of **4**, both absorption bands present significant hypsochromic and hyperchromic shifts compared to the free Schiff base, which can be attributed to the increased conjugation of the Cu(II)-Schiff base system. More specifically, the absorption band, indicative of the $\pi \to \pi^*$ transitions, appears at 274 nm whereas the absorption band that arises from the $n \to \pi^*$ transitions is located at 381 nm. No $d \to d$ transitions were observed in the spectrum of **4**, potentially because of the low concentration (10^{-5} M) of the solution. However, additional measurements at a higher concentration (10^{-3} M) (inset graph in Figure 2) showed a broad absorption band at 682 nm, indicative of $d \to d$ transitions.

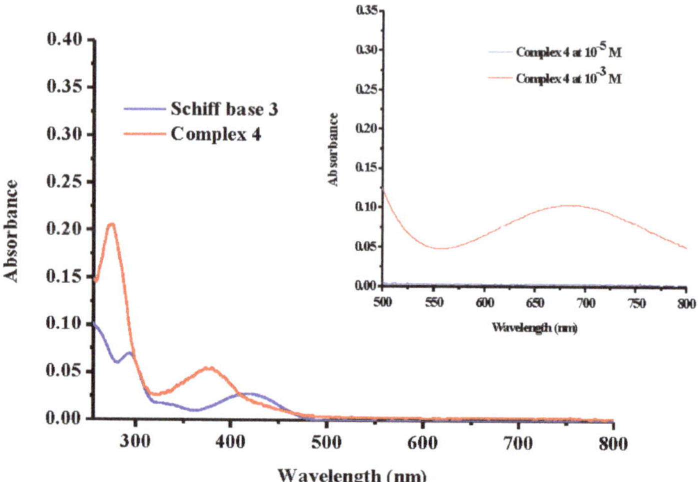

Figure 2. UV-Vis spectra of the Schiff base **3** (blue line) and complex **4** (red line) in MeOH at a concentration of 10^{-5} M. The inset graph presents the d-d transitions observed in MeOH at a concentration of 10^{-3} M.

3. Materials and Methods

The following starting materials were purchased from commercial sources (Sigma, Fluka, St. Louis, MO, USA) and were used without further purification: *Gamma*-aminobutyric acid, *ortho*-vanillin, copper nitrate trihydrate [$Cu(NO_3)_2 \cdot 3H_2O$], and sodium hydroxide (NaOH) pellets. Solvents: ethanol (EtOH), methanol (MeOH). The isolated and dried under vacuum at room temperature compounds are air-stable. Fourier transform-infrared (FT-IR) spectra were recorded on a Perkin Elmer 1760X spectrometer (Perkin-Elmer, San Francisco, CA, USA). A ThermoFinnigan Flash EA 1112 CHNS elemental analyzer (Waltham, MA, USA) was used for the simultaneous determination of carbon, hydrogen, and nitrogen (%). The analyzer operation is based on the dynamic flash combustion of the sample (at 1800 °C) followed by reduction, trapping, complete GC separation, and detection of the products. The instrument is fully automated and controlled by PC via the Eager 300 dedicated software (Thermo Fischer, Waltham, MA, USA). High resolution electrospray spray ionization mass spectra (HR-ESI-MS) of the Schiff base (**1**) and its Cu(II) complex (**2**) were obtained on UHPLC LC-MSn Orbitrap Velos-Thermo instrument (Thermo Scientific; Bremen, Germany) in the Institute of Biology, Medicinal Chemistry and Biotechnology of the National Hellenic Research Foundation. NMR spectra were recorded with a Bruker Avance 500 MHz spectrometer (Bruker, Rheinstetten, Germany) operating at 500 MHz (^{1}H) and 125 MHz (^{13}C). Chemical shifts are reported in ppm relative to DMSO-d_6 (^{1}H: δ = 2.50 ppm, ^{13}C: δ = 39.52 ± 0.06 ppm). NMR spectra assignments follow the numbering of the crystallographic analysis (Figure 1A). UV-Visible (UV-Vis) measurements were carried out on a Hitachi U2001 spectrophotometer (Hitachi, Tokyo, Japan) in the range from 200 to 800 nm.

3.1. Synthesis

3.1.1. (*E*)-4-[(2-Hydroxy-3-methoxybenzylidene)amino]butanoic Acid (**3**)

Ortho-vanillin **1** (152 mg, 1.0 mmol) dissolved in 0.5 mL MeOH, was added in a refluxing methanolic solution (6.5 mL) of *gamma*-aminobutyric acid **2** (103 mg, 1.0 mmol). Two drops of glacial acetic acid (0.1 mL, 1.75 mmol) were added and the resulting clear solution was stirred for 2 h. Upon completion of the reaction, the solvent was allowed to distill off by removing the reflux apparatus until reaching approximately 2/3 of its volume. Then, the solution was cooled to room temperature to afford the product as crystal needles. The bright yellow product (204 mg, 0.86 mmol, 86%) was filtered, washed once with a small volume of ice-cold MeOH and dried under vacuum. ^{1}H-NMR (500 MHz, DMSO-d_6): 1.86 (quint, J = 7.0 Hz, 2H, H-3), 2.29 (t, J = 7.0 Hz, 2H, H-2), 3.60 (t, J = 7.0 Hz, 2H, H-4), 3.77 (s, 3H, OMe), 6.79 (t, J = 7.9 Hz, 1H, H-10), 7.00 (t, J = 7.9 Hz, 1H, H-11), 7.02 (t, J = 7.9 Hz, 1H, H-9), 8.53 (s, 1H, H-5), 12.2 (br, 1H, COOH), 13.7 (br, 1H, OH) ppm; ^{13}C-NMR (125 MHz, DMSO-d_6): 25.8 (C-3), 31.2 (C-2), 55.7 (OMe), 57.0 (C-4), 114.7 (C-9), 117.5 (C-10), 118.2 (C-6), 123.1 (H-11), 148.1 (C-8), 151.9 (C-7), 166.2 (C-5), 174.0 (COOH) ppm; Anal. Calcd for **1**, $C_{12}H_{15}NO_4$: C, 60.75; H, 6.37; N, 5.90. Found: C, 60.69; H, 6.31; N, 5.94. HR-ESI-MS (positive mode), calcd. for $[(C_{12}H_{15}NO_4) + H]^+$ m/z = 238.1079, found m/z = 238.1076.

3.1.2. Synthesis of $[Cu(C_{12}H_{14}NO_4)_2]_n \cdot nO$ (4)

To a solution of *ortho*-vanillin **1** (0.15 g, 1.0 mmol) in MeOH (10 mL) *gamma*-aminobutyric acid **2** (0.10 g, 1.0 mmol) was added under stirring. The resulting yellow solution was refluxed for two hours at 60 °C under continuous stirring and then cooled to room temperature. Subsequently, a solution of $Cu(NO_3)_2 \cdot 3H_2O$ (0.12 g, 0.5 mmol) in MeOH (10 mL) was added under continuous stirring. The resulting clear, green reaction mixture was refluxed for an additional 2 h at 60 °C and then cooled to room temperature. To that, NaOH (0.04 g, 1 mmol) was added under continuous stirring. The resulting homogeneous dark green reaction mixture was refluxed at 60 °C for an additional 2 h and then cooled to room temperature. Subsequently, the reaction flask was left to evaporate slowly at room temperature. One week later, dark green plate-like crystalline material precipitated at the bottom of the flask. The product was isolated by filtration and dried in vacuo. Yield: 0.14 g (52%). Anal. Calcd for **2**, $[Cu(C_{12}H_{14}NO_4)_2]_n \cdot nO$. ($C_{24}H_{28}CuN_2O_9$, M_r 540.04): C, 53.38; H, 5.23; N, 5.19. Found: C, 53.36; H, 5.18; N, 5.14. HR-ESI-MS (positive mode), calcd, for $\{[Cu(C_{12}H_{14}NO_4)_2] + H\}^+$ m/z = 536.1219, found m/z = 536.1215.

3.2. X-ray Crystal Structure Determination

X-ray quality crystals of **4** were grown from MeOH. Crystals of **4** suitable for X-ray diffraction, with dimensions 0.14 × 0.09 × 0.05 mm were taken from the mother liquor and mounted at room temperature on a Bruker Kappa APEX 2 diffractometer (Bruker AXS, Madison, WI, USA), equipped with a triumph monochromator, using Mo Kα radiation. Cell dimensions and crystal system determination were performed using 172 high θ reflections with $10° < \theta < 20°$. Data collection (φ- and ω- scans) and processing (cell refinement, data reduction and numerical absorption correction based on dimensions) were performed using the SAINT and SADABS programs [23,24]. The structure was solved by the SUPERFLIP package [25]. The CRYSTALS version 14.61 build 6236 program package was used for structure refinement (full-matrix least-squares methods on F^2) and all subsequently remaining calculations [26]. Molecular illustrations were drawn using the CAMERON crystallographic package [27]. All non-hydrogen non-disordered atoms were anisotropically refined. All hydrogen atoms were found at their expected positions and were refined using proper riding constraints to the pivot atoms. Crystallographic details for **4** are summarized in Table S3 (Supplementary Information). Further details on the crystallographic studies as well as atomic displacement parameters are given as Supporting Information as well as in the form of cif file.

4. Conclusions

Two novel compounds, the Schiff base (*E*)-4-[(2-hydroxy-3-methoxybenzylidene)-amino]butanoic acid and its Cu(II) complex were prepared in good yields and purity. The compounds have been physico-chemically and structurally characterized via elemental analysis, HR-ESI-MS, FT-IR, UV-Vis, NMR, and single crystal X-ray diffraction. The crystal structure of the produced Cu(II) complex reflects an one-dimensional polymeric compound. The Cu(II) ion is bound to two singly deprotonated Schiff base bridging ligands forming a $Cu^{II}N_2O_4$ chelation environment, and a coordination sphere with a disordered octahedral geometry. Further investigation is underway to determine the biological activities of a library of analogue derivatives.

Supplementary Materials: The following are available online. Table S1. Hydrogen bonds in **4**. Table S2: Summary of crystal, intensity collection and refinement data for $[Cu(C_{12}H_{14}NO_4)_2]_n \cdot nO$ (**4**). Figure S1. ^{1}H-NMR spectrum of compound **3** in DMSO-d_6. Figure S2. ^{13}C-NMR spectrum of compound **3** in DMSO-d_6. Figure S3. ^{1}H-^{1}H COSY spectrum of compound **3** in DMSO-d_6. Figure S4. ^{1}H-^{13}C HMBC spectrum of compound **3** in DMSO-d_6. Figure S5. ^{1}H-^{13}C HSQC spectrum of compound **3** in DMSO-d_6. Figure S6. ^{1}H-^{1}H NOESY spectrum of compound **3** in DMSO-d_6. Figure S7. FT-IR spectra of Schiff base **3** and complex **4**. Figure S8. HRMS Spectra of the Schiff base **3**. Figure S9. HRMS Spectrum of the complex **4**. CCDC 2051769 (4) contains the supplementary crystallographic data for this paper. These data can be obtained free of charge via www.ccdc.cam. ac.uk/conts/retrieving.html (or from the Cambridge Crystallographic Data Centre, 12 Union Road, Cambridge CB21EZ, UK; fax: (+44)-1223-336-033; or deposit@ccde.cam.ac.uk).

Author Contributions: E.H.: inorganic synthesis, original draft preparation, methodology, metal complex structural elucidation; A.H.: X-ray crystallographic analysis; B.M.: FT-IR spectroscopic analysis; M.S.: review, editing, investigation; M.P.: review, editing, investigation; D.M.: organic synthesis, original draft preparation, methodology, ligand structural elucidation. All authors have read and agreed to the published version of the manuscript.

Funding: E. Halevas gratefully acknowledges financial support by Stavros Niarchos Foundation (SNF) through implementation of the program of Industrial Fellowships at NCSR "Demokritos" and the Foundation for Education and European Culture (IPEP) founded by Nicos and Lydia Tricha. B.M. gratefully acknowledges the financial support by the State Scholarships Foundation (IKΥ) through the implementation of the program of Industrial Fellowships at NCSR "Demokritos" and "Reinforcement of Postdoctoral Researchers—2nd Cycle"(MIS-5033021) (European Social Fund (ESF)—Operational Programme "Human Resources Development, Education and Lifelong Learning".

Data Availability Statement: The data presented in this study are available in this article and Supplementary Materials.

Conflicts of Interest: The authors declare no conflict of interest.

References

1. Kostova, I.; Saso, L. Advances in research of Schiff-base metal complexes as potent antioxidants. *Curr. Med. Chem.* **2013**, *20*, 4609–4632. [CrossRef]
2. Ejidike, I.P.; Ajibade, P.A. Transition metal complexes of symmetrical and asymmetrical Schiff bases as antibacterial, antifungal, antioxidant, and anticancer agents: Progress and prospects. *Rev. Inorg. Chem.* **2015**, *35*, 191–224. [CrossRef]
3. Halevas, E.; Tsave, O.; Yavropoulou, M.P.; Hatzidimitriou, A.; Yovos, J.G.; Psycharis, V.; Gabriel, C.; Salifoglou, A. Design, synthesis and characterization of novel binary V(V)-Schiff base materials linked with insulin-mimetic vanadium-induced differentiation of 3T3-L1 fibroblasts to adipocytes. Structure–function correlations at the molecular level. *J. Inorg. Biochem.* **2015**, *147*, 99–115. [CrossRef] [PubMed]
4. Halevas, E.; Nday, C.M.; Chatzigeorgiou, E.; Varsamis, V.; Eleftheriadou, D.; Jackson, G.E.; Litsardakis, G.; Lazari, D.; Ypsilantis, K.; Salifoglou, A. Chitosan encapsulation of essential oil "cocktails" with well-defined binary Zn(II)-Schiff base species targeting antibacterial medicinal nanotechnology. *J. Inorg. Biochem.* **2017**, *176*, 24–37. [CrossRef] [PubMed]
5. Halevas, E.; Tsave, O.; Yavropoulou, M.; Yovos, J.G.; Hatzidimitriou, A.; Psycharis, V.; Salifoglou, A. In vitro structure-specific Zn(II)-induced adipogenesis and structure-function bioreactivity correlations. *J. Inorg. Biochem.* **2017**, *177*, 228–246. [CrossRef] [PubMed]
6. Shamsi, M.; Yadav, S.; Arjmand, F. Synthesis and characterization of new transition metal Cu(II), Ni(II) and Co(II) L phenylalanine-DACH conjugate complexes: In vitro DNA binding, cleavage and molecular docking studies. *J. Photochem. Photobiol. B* **2014**, *136*, 1–11. [CrossRef]
7. Fattuoni, C.; Vascellari, S.; Pivetta, T. Synthesis, protonation constants and biological activity determination of amino acid–salicylaldehyde-derived Schiff bases. *Amino Acids* **2020**, *52*, 397–407. [CrossRef]
8. Gupta, K.C.; Kumar, A.S.; Lin, C.C. Polymer-supported Schiff base complexes in oxidation reactions. *Coord. Chem. Rev.* **2009**, *253*, 1923–1946. [CrossRef]
9. Abdel-Rahman, L.H.; El-Khati, R.M.; Nassr, L.A.E.; Abu-Dief, A.M. DNA binding ability mode, spectroscopic studies, hydrophobicity, and in vitro antibacterial evaluation of some new Fe(II) complexes bearing ONO donors amino acid Schiff bases. *Arab. J. Chem.* **2017**, *10*, S1835–S1846. [CrossRef]
10. Da Silva, C.M.; da Silva, D.L.; Modolo, L.V.; Alves, R.B.; de Resende, M.A.; Martins, C.V.B.; de Fátima, Â. Schiff bases: A short review of their antimicrobial activities. *J. Adv. Res.* **2011**, *2*, 1–8. [CrossRef]
11. Li, Y.; Dong, J.; Zhao, P.; Hu, P.; Yang, D.; Gao, L.; Li, L. Synthesis of amino acid Schiff base nickel (II) complexes as potential anticancer drugs in vitro. *Bioinorg. Chem. Appl.* **2020**, *2020*, 8834859. [CrossRef] [PubMed]

12. Abu-Dief, A.M.; Mohamed, I.M.A. A review on versatile applications of transition metal complexes incorporating Schiff bases. *Beni-Suef Univ. J. Basic Appl. Sci.* **2015**, *4*, 119–133. [CrossRef] [PubMed]
13. Zou, Y.; Han, J.; Saghyan, A.S.; Mkrtchyan, A.F.; Konno, H.; Moriwaki, H.; Izawa, K.; Soloshonok, V.A. Asymmetric synthesis of tailor-made amino acids using chiral Ni(II) complexes of Schiff bases. An update of the recent literature. *Molecules* **2020**, *25*, 2739. [CrossRef]
14. Puterová-Tokárová, Z.; Mrázová, V.; Boča, R. Magnetism of novel Schiff-base copper(II) complexes derived from aminoacids. *Polyhedron* **2013**, *61*, 87–93. [CrossRef]
15. Wang, Y.-Y.; Xu, F.-Z.; Zhu, Y.-Y.; Song, B.; Luo, D.; Yu, G.; Chen, S.; Xue, W.; Wu, J. Pyrazolo[3,4-d]pyrimidine derivatives containing a Schiff base moiety as potential antiviral agents. *Bioorg. Med. Chem. Lett.* **2018**, *28*, 2979–2984. [CrossRef]
16. Jamil, D.M.; Al-Okbi, A.K.; Al-Baghdadi, S.B.; Al-Amiery, A.A.; Kadhim, A.; Gaaz, T.S.; Kadhum, A.A.H.; Mohamad, A.B. Experimental and theoretical studies of Schiff bases as corrosion inhibitors. *Chem. Cent. J.* **2018**, *12*, 7. [CrossRef]
17. Mobinikhaledi, A.; Jabbarpour, M.; Hamta, A. Synthesis of some novel and biologically active Schiff bases bearing a 1,3,4-thiadiazole moiety under acidic and PTC conditions. *J. Chil. Chem. Soc.* **2011**, *56*, 812–814. [CrossRef]
18. Galić, N.; Cimerman, Z.; Tomišić, V. Spectrometric study of tautomeric and protonation equilibria of o-vanillin Schiff base derivatives and their complexes with Cu(II). *Spectrochim. Acta A Mol. Biomol. Spectrosc.* **2008**, *71*, 1274–1280. [CrossRef]
19. Pis-Diez, R.; Echeverria, G.A.; Piro, O.E.; Jios, J.L. Parajón-Costa, B.S. A structural, spectroscopic and theoretical study of an o-vanillin Schiff base derivative involved in enol-imine and keto-amine tautomerism. *New J. Chem.* **2016**, *40*, 2730–2740. [CrossRef]
20. Muche, S.; Harms, K.; Biernasiuk, A.; Malm, A.; Popiołek, Ł.; Hordyjewska, A.; Olszewska, A.; Hołyńska, M. New Pd(II) Schiff base complexes derived from ortho-vanillin and L-tyrosine or L-glutamic acid: Synthesis, characterization, crystal structures and biological properties. *Polyhedron* **2018**, *151*, 465–477. [CrossRef]
21. Halevas, E.; Pekou, A.; Papi, R.; Mavroidi, B.; Hatzidimitriou, A.G.; Zahariou, G.; Litsardakis, G.; Sagnou, M.; Pelecanou, M.; Pantazaki, A.A. Synthesis, physicochemical characterization and biological properties of two novel Cu(II) complexes based on natural products curcumin and quercetin. *J. Inorg. Biochem.* **2020**, *208*, 111083. [CrossRef] [PubMed]
22. Munde, A.S.; Jagdale, A.N.; Jadhav, S.M.; Chondhekar, T.K. Synthesis, characterization and thermal study of some transition metal complexes of an asymmetrical tetradentate Schiff base ligand. *J. Serb. Chem. Soc.* **2010**, *75*, 349–359. [CrossRef]
23. Bruker Analytical X-Ray Systems, Inc. *Apex2, Version 2 User Manual: M86-E01078*; Bruker Analytical X-Ray Systems, Inc.: Madison, WI, USA, 2006.
24. Siemens Industrial Automation, Inc. *SADABS: Area-Detector Absorption Correction*; Siemens Industrial Automation, Inc.: Madison, WI, USA, 1996.
25. Betteridge, P.W.; Carruthers, J.R.; Cooper, R.I.; Prout, K.; Watkin, D.J. CRYSTALS version 12: Software for guided crystal structure analysis. *J. Appl. Cryst.* **2003**, *36*, 1487. [CrossRef]
26. Palatinus, L.; Chapuis, G. SUPERFLIP–a computer program for the solution of crystal structures by charge flipping in arbitrary dimensions. *Appl. Cryst.* **2007**, *40*, 786–790. [CrossRef]
27. Watkin, D.J.; Prout, C.K.; Pearce, L.J. *CAMERON*; Chemical Crystallography Laboratory: Oxford, UK, 1996.

Short Note

Cis-Bis(2,2′-Azopyridinido)dicarbonylruthenium(II)

Tsugiko Takase [1],*, Shuya Kainuma [2], Takatoshi Kanno [2] and Dai Oyama [1],*

[1] Department of Natural Sciences and Informatics, Fukushima University, 1 Kanayagawa, Fukushima 960-1296, Japan

[2] Graduate School of Science and Engineering, Fukushima University, 1 Kanayagawa, Fukushima 960-1296, Japan; s.016.kai@gmail.com (S.K.); s1970013@ipc.fukushima-u.ac.jp (T.K.)

* Correspondence: ttakase@sss.fukushima-u.ac.jp (T.T.); daio@sss.fukushima-u.ac.jp (D.O.); Tel.: +81-24-548-8199 (D.O.)

Abstract: An [Ru(apy)$_2$Cl$_2$] precursor (apy = 2,2′-azopyridine) in 2-methoxyethanol was heated under a pressurized CO atmosphere to afford a diradical complex, [Ru(apy·$^-$)$_2$(CO)$_2$], containing one-electron-reduced azo anion radical ligands. The electronic states of the complex were characterized by spectroscopic techniques and computational studies. Magnetic measurements revealed the existence of antiferromagnetic interactions in the diradical complex.

Keywords: ruthenium; carbonyl complex; azopyridine; anion radical; electronic structure; magnetic properties

Citation: Takase, T.; Kainuma, S.; Kanno, T.; Oyama, D. *Cis*-Bis(2,2′-Azopyridinido)dicarbonylruthenium (II). *Molbank* **2021**, *2021*, M1182. https://doi.org/10.3390/M1182

Academic Editors: Dimitrios Matiadis and Eleftherios Halevas

Received: 28 December 2020

Accepted: 15 January 2021

Published: 18 January 2021

Publisher's Note: MDPI stays neutral with regard to jurisdictional claims in published maps and institutional affiliations.

1. Introduction

The oxidation number of the central metal atom in a complex generally determines the electronic states of the complex. In a metal complex containing a redox-active non-innocent ligand, conversely, ligands as well as the metal centers can also control the electronic states of the complex. For example, azo compounds can directly accept one or two electrons because they have a low-lying azo-centered vacant π* molecular orbital (Scheme 1) [1]. Due to this property, many transition metal complexes, particularly those in groups 7 and 8, containing azopyridyl ligands, e.g., 2,2′-azopyridine (apy) and 2-phenylazopyridine (pap), (Figure 1a), have been reported [2–7]. Based on numerous studies, metal complexes containing non-innocent azo ligands are expected to be applied to multifunctional materials which surpass conventional magnetic materials, such as metal oxides [8,9].

Scheme 1. Two-step electron transfers in azo compounds.

Figure 1. Chemical structures of (**a**) azopyridines; (**b**) The Ru-complex presented in this study.

We have previously reported the synthesis and properties of a monocarbonylruthenium(II) complex containing two azopyridine molecules [10]. All the azopyridyl ligands in the monocarbonyl complex were neutral (nonradical, singlet state: $S = 0$). In contrast, the newly synthesized dicarbonylruthenium(II) complex in this work (Figure 1b) experimentally and theoretically revealed that both the azopyridyl ligands are coordinated as anion radicals (triplet state: $S = 1$). Additionally, the magnetic properties of the diradical complex were examined.

2. Results and Discussion

2.1. Synthesis and Characterization of the Diradical Complex

We previously reported the detailed characterization, including X-ray crystallography, of the monocarbonyl complex ([Ru(pap)$_2$(CO)Cl]$^+$) [10]. In this report, we confirmed that the two azo ligands were neutral in the monocarbonyl complex. Meanwhile, two carbonyl stretching frequencies in the newly prepared dicarbonyl complex were observed at 2060 and 1995 cm^{-1} in the IR spectra (Figure S1). These values are 30–40 cm^{-1} lower than those of similar Ru complexes ([Ru(N-N)$_2$(CO)$_2$]$^{2+}$; N-N = bidentate pyridyl ligands) [11–13]. The reason is that the reduction of the azo moiety increases the electron density of the complex, resulting in a red shift of the CO bands [5]. Additionally, no peaks assignable to the present complex were observed in the mass spectra, and its ^{1}H-NMR spectrum was silent. It is, therefore, interpreted that the introduction of two CO molecules, which function as ligand and reducing agents, led to the selective formation of the diradical neutral complex, in which both the azo ligands underwent one-electron reduction (Equation (1)).

$$[\text{Ru}^{\text{II}}(\text{apy})_2\text{Cl}_2] + 2\text{CO} \rightarrow [\text{Ru}^{\text{II}}(\text{apy}\cdot^-)_2(\text{CO})_2] + 2\text{Cl}^- \tag{1}$$

2.2. Electronic Structure Analysis

We performed quantum chemical calculations for both the diradical and the corresponding nonradical complexes to evaluate their stability (Figure S2). The adiabatic energy gap between the two complexes in methanol were calculated (1530.0478 Hartree for the nonradical state and 1530.3652 Hartree for the diradical one) [14]. The results indicate that the diradical state is 30.75 kcal/mol more stable than the nonradical one. Figure 2 shows the spin densities in the optimized structures. Since the spin densities of both azo sites in the diradical state are 0.623, the azo site is the primary spin-bearing center.

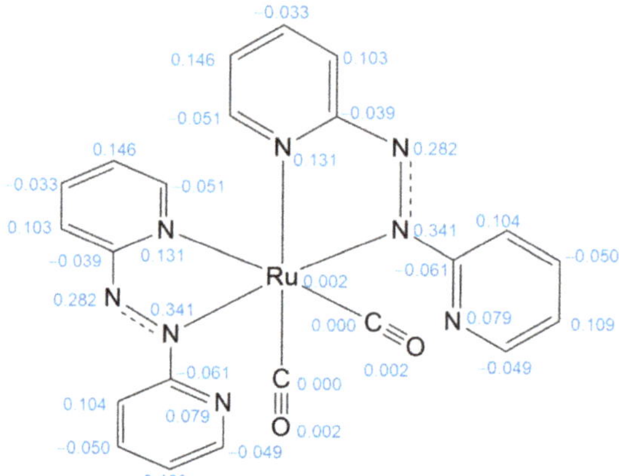

Figure 2. Spin density analyses for [Ru(apy·$^-$)$_2$(CO)$_2$], calculated at the DFT/B3LYP/LanL2DZ/6-31G(d) level of theory.

Figure 3a shows the estimated absorption spectra of the nonradical ($S = 0$) and diradical complexes ($S = 1$) using time-dependent density functional theory (TDDFT) calculations. It is presumed that the nonradical complex exhibits absorption maximum around 400 nm, while the diradical one exhibits absorptions around 350 and 470 nm. Since the latter values are consistent with the spectral data of the prepared complex (λ_{max} = 355 and 487 nm; Figure 3b), computational results also suggested that the present complex contains two anion radical ligands (apy$\cdot^{-}$).

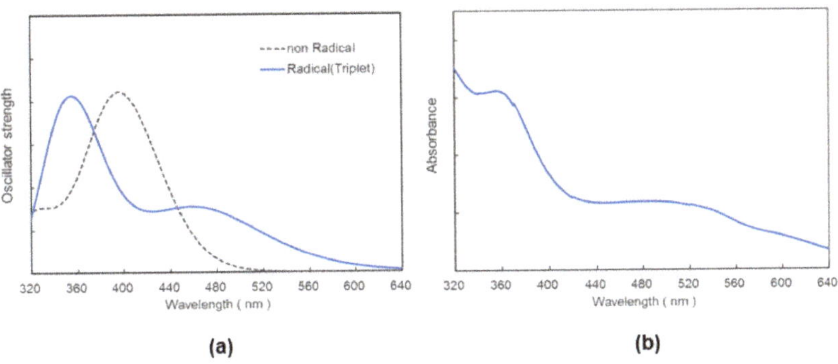

Figure 3. (a) The theoretical absorption spectra calculated for the singlet and triplet spin configurations of [Ru(apy)$_2$(CO)$_2$]$^{2+/0}$ at the TDDFT/B3LYP/LanL2DZ/6-31G(d) level of theory; (b) The absorption spectrum of [Ru(apy$\cdot^{-}$)$_2$(CO)$_2$].

2.3. Magnetic Properties

To experimentally confirm that the present complex is a radical complex with unpaired spins, we carried out measurements of the dependence of the magnetic moment on the external magnetic field (*M-H* properties). Since $M = \chi H$ (*M*: magnetic moment, *H*: external applied magnetic field, χ: magnetic molar susceptibility) is satisfied within the temperature range measured, the complex is the radical complex, which displays paramagnetic properties. Additionally, the Curie–Weiss law defined by Equation (2) (*C*: Curie constant, *T*: absolute temperature, θ: Weiss constant) is satisfied in the low-temperature region bearing on the temperature dependence of the magnetic susceptibility (the applied magnetic field of 10000 Oe) as shown in Figure 4; the Weiss constant is negative ($\theta = -10.7$ K), suggesting that antiferromagnetic interactions exist between the spins in each radical [5].

$$\chi = C/(T - \theta) \tag{2}$$

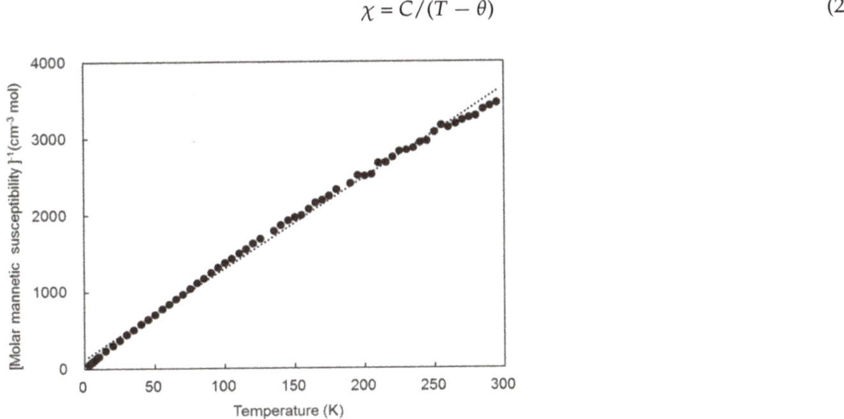

Figure 4. Temperature dependence of the inverse molar magnetic susceptibility for [Ru(apy$\cdot^{-}$)$_2$(CO)$_2$].

In summary, we have found that the electronic state of an azo ligand can be controlled by the number of coordinated carbonyls, i.e., the monocarbonyl complex has neutral azo ligands, whereas the corresponding dicarbonyl one led to anion radicals. The latter complex exhibited paramagnetic and antiferromagnetic interactions between the spins in each radical. Further studies using analogous compounds are underway, directed toward the development of multifunctional materials in detail.

3. Materials and Methods

3.1. Measurements

The infrared (IR) spectra were obtained using a JASCO FT-IR 4100 spectrometer (JASCO Corporation, Hachioji, Japan). The elemental analysis data were obtained on a Perkin-Elmer 2400II series CHN analyzer (Perkin-Elmer, Inc., Waltham, MA, USA). The absorption spectra were obtained on a JASCO V-560 spectrophotometer (JASCO Corporation, Hachioji, Japan). The magnetic susceptibilities of the complex were recorded on a SQUID magnetometer (MPMS, Quantum Design, Inc., San Diego, CA, USA) in the temperature range of 5–300 K. DFT and TDDFT calculations were performed using the quantum chemical program, Gaussian 16 [15]. The geometries of the complex in its singlet ($S = 0$) or triplet ($S = 1$) states were fully optimized by restricted (singlet) or unrestricted (triplet) DFT methods employing the B3LYP function [16,17], with a 6-31G(d) basis set for the light elements [18,19] and a LanL2DZ basis set [20] for the Ru atom. The solvent effect of methanol was evaluated using an implicit solvent model, and a conductor-like polarizable continuum model (CPCM). Vibrational analyses were performed at the same calculation level employed for geometry optimization. The spin density analyses were also calculated at the optimized geometry. Excited-state calculations based on the B3LYP-optimized geometries were conducted for the TDDFT formalism in methanol using the CPCM.

3.2. Synthesis of the Complex

The compound, 2,2′-azopyridine (apy), and the starting material ([Ru(apy)$_2$Cl$_2$]) were prepared according to known procedures or through modification of published methods [21,22].

A solution of [Ru(apy)$_2$Cl$_2$] (38 mg, 0.071 mmol) in 2-methoxyethanol (40 mL) was heated to 100 °C under an atmosphere of CO (2 MPa) for 72 h. The volume was reduced to 5 mL using a rotary evaporator, and a black precipitate was formed by the addition of diethyl ether. The product was collected by filtration, washed with diethyl ether, and dried in vacuo. The crude product was recrystallized from dichloromethane and diethyl ether. Yield: 33 mg (88%). Anal. Calc. for $C_{22}H_{16}N_8O_2Ru \cdot 2.5CH_2Cl_2$: C, 39.88; H, 2.87; N, 15.19. Found: C, 39.82; H, 2.89; N, 15.28%. IR (KBr): $\nu = 2060$ and 1995 cm^{-1} (C≡O). UV-vis (dimethyl sulfoxide): $\lambda_{max}/$nm ($\varepsilon/$M^{-1} cm^{-1}) 355 (6800) and 487 (2900).

Supplementary Materials: The following are available online, Figure S1: The IR spectrum of [Ru(apy·$^-$)$_2$(CO)$_2$] (KBr method), Figure S2: Optimized structures of [Ru(apy·$^-$)$_2$(CO)$_2$] and [Ru(apy)$_2$(CO)$_2$]$^{2+}$ (in MeOH).

Author Contributions: Conceptualization, T.T. and D.O.; synthesis, S.K.; formal analysis, S.K. and T.K.; investigation, T.T.; writing—original draft preparation, T.T.; writing—review and editing, D.O.; supervision, D.O.; funding acquisition, D.O. All authors have read and agreed to the published version of the manuscript.

Funding: This research was funded by JSPS KAKENHI, grant numbers JP17K05799 and JP20K05536.

Data Availability Statement: The data presented in this study are available in this article and supporting supplementary material.

Conflicts of Interest: The authors declare no conflict of interest.

References

1. Kaim, W. Complexes with 2,2′-azobispyridine and related 'S-frame' bridging ligands containing the azo function. *Coord. Chem. Rev.* **2001**, *219–221*, 463–488. [CrossRef]
2. Shivakumar, M.; Pramanik, K.; Bhattacharyya, I.; Chakravorty, A. Chemistry of metal-bound anion radicals. A family of mono- and bis(azopyridine) chelates of bivalent ruthenium. *Inorg. Chem.* **2000**, *39*, 4332–4338. [CrossRef] [PubMed]
3. Das, D.; Mondal, T.K.; Mobin, S.M.; Lahiri, G.K. Sensitive valence structures of [(pap)$_2$Ru(Q)]^n (n = +2, +1, 0, −1, −2) with two different redox noninnocent ligands, Q = 3,5-di-*tert*-butyl-*N*-aryl-1,2-benzoquinonemonoimine and pap = 2-phenylazopyridine. *Inorg. Chem.* **2009**, *48*, 9800–9810. [CrossRef] [PubMed]
4. Oyama, D.; Asuma, A.; Hamada, T.; Takase, T. Novel [Ru(polypyridine)(CO)$_2$Cl$_2$] and [Ru(polypyridine)$_2$(CO)Cl]$^+$-type complexes: Characterizing the effects of introducing azopyridyl ligands by electrochemical, spectroscopic and crystallographic measurements. *Inorg. Chim. Acta* **2009**, *362*, 2581–2588. [CrossRef]
5. Paul, N.; Samanta, S.; Goswami, S. Redox induced electron transfer in doublet azo-anion diradical rhenium(II) complexes. Characterization of complete electron transfer series. *Inorg. Chem.* **2010**, *49*, 2649–2655. [CrossRef]
6. Ghosh, P.; Samanta, S.; Roy, S.K.; Demeshko, S.; Meyer, F.; Goswami, S. Introducing a new azoaromatic pincer ligand. Isolation and characterization of redox events in its ferrous complexes. *Inorg. Chem.* **2014**, *53*, 4678–4686. [CrossRef]
7. Oyama, D.; Mun, B.; Takase, T. Redox-induced reversible intramolecular carbon-nitrogen bond formation of an azopyridylruthenium complex: Control of carbonyl ligand photoreactivity caused by structural change of the complex. *J. Organomet. Chem.* **2015**, *799–800*, 173–178. [CrossRef]
8. Zapata-Rivera, J.; Maynau, D.; Calzado, C.J. Evaluation of the magnetic interactions in salts containing [Ni(dmit)$_2$]$^-$ radical anions. *Chem. Mater.* **2017**, *29*, 4317–4329. [CrossRef]
9. Farcaş, A.-A.; Beu, T.A.; Bende, A. Light-induced spin transitions in Ni(II)-based macrocyclic-ligand complexes: A DFT study. *J. Photochem. Photobiol.* **2019**, *376*, 316–323. [CrossRef]
10. Oyama, D.; Takatsuki, Y.; Fujita, R. Azopyridylruthenium(II) complexes containing Ru-C bonds: Synthesis, characterization and reactivity. *Trends Inorg. Chem.* **2010**, *12*, 31–40.
11. Kobayashi, K.; Tanaka, K. Reactivity of CO$_2$ activated on transition metals and sulfur ligands. *Inorg. Chem.* **2015**, *54*, 5085–5095. [CrossRef] [PubMed]
12. Oyama, D.; Abe, R.; Takase, T. CO-ligand photodissociation in two Ru(II) complexes affected by different polypyridyl supporting ligands. *Chem. Lett.* **2017**, *46*, 1412–1414. [CrossRef]
13. Akatsuka, K.; Abe, R.; Takase, T.; Oyama, D. Coordination chemistry of Ru(II) complexes of an asymmetric bipyridine analogue: Synergistic effects of supporting ligand and coordination geometry on reactivities. *Molecules* **2020**, *25*, 27. [CrossRef]
14. Luckhaus, D.; Yamamoto, Y.-I.; Suzuki, T.; Signorell, R. Genuine binding energy of the hydrated electron. *Sci. Adv.* **2017**, *3*, e1603224. [CrossRef]
15. Frisch, M.J.; Trucks, G.W.; Schlegel, H.B.; Scuseria, G.E.; Robb, M.A.; Cheeseman, J.R.; Scalmani, G.; Barone, V.; Petersson, G.A.; Nakatsuji, H.; et al. *Gaussian 16 (Revision, A.03)*; Gaussian, Inc.: Wallingford, CT, USA, 2016.
16. Becke, A.D. Density-functional thermochemistry. III. The role of exact exchange. *J. Chem. Phys.* **1993**, *98*, 5648–5652. [CrossRef]
17. Lee, C.; Yang, W.; Parr, R.G. Development of the Colle-Salvetti correlation-energy formula into a functional of the electron density. *Phys. Rev.* **1988**, *B37*, 785–789. [CrossRef]
18. Hehre, W.J.; Ditchfield, R.; Pople, J.A. Self-consistent molecular orbital methods. XII. Further extensions of Gaussian-type basis sets for use in molecular orbital studies of organic molecules. *J. Chem. Phys.* **1972**, *56*, 2257–2261. [CrossRef]
19. Francl, M.M.; Pietro, W.J.; Hehre, W.J.; Binkley, J.S.; Gordon, M.S.; DeFrees, D.J.; Pople, J.A. Self-consistent molecular orbital methods. XXIII. A polarization-type basis set for second-row elements. *J. Chem. Phys.* **1982**, *77*, 3654–3665. [CrossRef]
20. Wadt, W.R.; Hay, P.J. Ab initio effective core potentials for molecular calculations. Potentials for main group elements Na to Bi. *J. Chem. Phys.* **1985**, *82*, 284–298. [CrossRef]
21. Baldwin, D.A.; Lever, A.B.P.; Parish, R.V. Complexes of 2,2′-azopyridine with iron(II), cobalt(II), nickel(II), copper(I), and copper(II). Infrared study. *Inorg. Chem.* **1969**, *8*, 107–115. [CrossRef]
22. Hotze, A.C.G.; Caspers, S.E.; DeVos, D.; Kooijman, H.; Spek, A.L.; Flamigni, A.; Bacac, M.; Sava, G.; Haasnoot, J.G.; Reedijk, J. Structure-dependent in vitro cytotoxicity of the isomeric complexes [Ru(L)$_2$Cl$_2$] (L=*o*-tolylazopyridine and 4-methyl-2-phenylazopyridine) in comparison to [Ru(azpy)$_2$Cl$_2$]. *J. Biol. Inorg. Chem.* **2004**, *9*, 354–364. [CrossRef]

Short Note

Homopiperazine (Hexahydro-1,4-diazepine)

R. Alan Aitken *, Dheirya K. Sonecha and Alexandra M. Z. Slawin

EaStCHEM School of Chemistry, University of St Andrews, North Haugh, St Andrews, Fife, KY16 9ST, UK;
dks1@st-and.ac.uk (D.K.S.); amzs@st-and.ac.uk (A.M.Z.S.)
* Correspondence: raa@st-and.ac.uk; Tel.: +44-1334-463865

Abstract: The X-ray structure of the title compound has been determined for the first time. Data on its ^{1}H–^{13}C-NMR coupling constants and ^{15}N-NMR spectrum are also given.

Keywords: homopiperazine; X-ray structure; C–C bond length; ^{15}N-NMR

check for updates

Citation: Aitken, R.A.; Sonecha, D.K.; Slawin, A.M.Z. Homopiperazine (Hexahydro-1,4-diazepine). *Molbank* **2021**, *2021*, M1200. https://doi.org/10.3390/M1200

Academic Editors: Dimitrios Matiadis and Eleftherios Halevas

Received: 30 March 2021
Accepted: 9 April 2021
Published: 10 April 2021

Publisher's Note: MDPI stays neutral with regard to jurisdictional claims in published maps and institutional affiliations.

1. Introduction

The fundamental heterocyclic compound hexahydro-1,4-diazepine or "homopiperazine" **1** was first reported in 1899 [1,2] and a convenient industrial preparation was described in 1961 [3]. At present, it is commercially available from several major suppliers and it has recently found various applications, including as a component of liquids for CO_2 capture [4] and as a component of various organic and organic/inorganic supramolecular ionic salts and transition metal complexes. In these latter applications, the materials are often characterised by X-ray diffraction and so there are a large number of X-ray structures involving a homopiperazinium dication in salts or mixed salts with, for example, carboxylic acids [5–7], potassium perchlorate [8], ammonium perchlorate [9], manganese sulfate oxalate [10], cobalt sulfate [11], zinc phosphate [12], arsenic oxide [13], tellurium phosphate [14], cadmium phosphate [15], cadmium chlorides [16,17], lead bromide [18], bismuth iodide [19], uranium fluoride [20] and uranium sulfate [21]. There are also a significant number of structures in which homopiperazine acts as a bidentate ligand on nickel [22,23], copper [24] and platinum [25–28]. However, in all of these, the structure is significantly distorted either by protonation or metal complexation, and it appears that the X-ray structure of homopiperazine on its own has not so far been determined, quite possibly because its low melting point of 42 °C [2] makes it difficult to obtain suitable crystals. We have now determined its structure and present here the result as well as several key NMR parameters.

2. Results

Crystals suitable for diffraction were obtained from the inner surface of a commercial bottle (Sigma Aldrich) which had been stored for >5 years, allowing slow sublimation. The resulting structure (Figure 1) shows how three adjacent molecules are aligned with the NH atoms of the central ring, viewed from above, hydrogen-bonding to the nitrogen lone pairs of the molecules on either side, which are viewed side-on. The bond lengths and angles in the symmetrical molecule are unremarkable, with the exception of the carbon–carbon bond of the N–CH$_2$–CH$_2$–N moiety, which is rather long at 1.603(10) Å. This may be compared with a mean value of 1.524 Å for sp^3 CH$_2$–CH$_2$ in general [29] and is in stark contrast to the values of 1.5204(15) and 1.5190(15) Å observed in piperazine hydrate [30], the value of 1.513(2) Å in the monoprotonated homopiperazinium perchlorate [31] and 1.521(4) Å in an *N,N'*-dibenzylatedhomopiperazine [32]. The reason for the long bond in this case is unclear, although other examples of such long NCH$_2$–CH$_2$N bonds have been located [33].

109

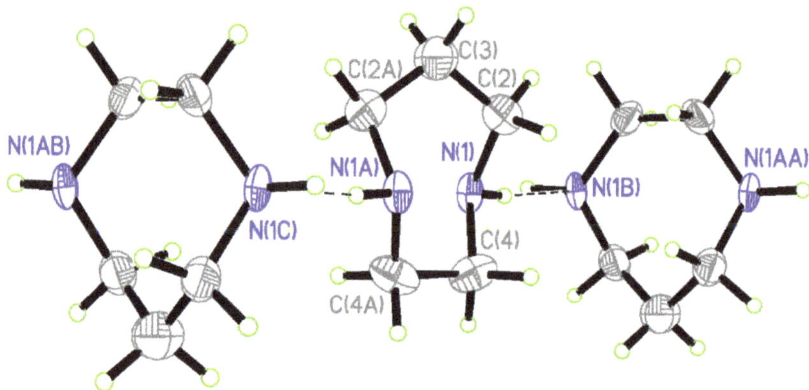

Figure 1. Molecular structure of **1** with anisotropic displacement ellipsoids drawn at 50% probability level and showing numbering system used. Bond lengths and angles: N(1)–C(2) 1.455(5), C(2)–C(3) 1.481(5), N(1)–C(4) 1.470(5), C(4)–C(4A) 1.603(10), N(1)–H(1) 0.977(7) Å; C(2)–N(1)–C(4) 112.0(3), N(1)–C(4)–C(4A) 111.0(3), N(1)–C(2)–C(3) 114.9(4), C(2)–C(3)–C(2A) 114.8(6)°. Symmetry operator to generate the whole molecule is −X + 1, −Y, Z.

The conformation of **1** is shown in Figure 2 and the N–C–C–N torsion angle of 58.2(4)° confirms that the left-hand end of the molecule has an almost ideal chair-type conformation, which is only slightly distorted by the insertion of the extra atom at the right-hand end. The two nitrogen lone pairs are aligned in opposite directions, which facilitates the formation of a hydrogen-bonded network due to interaction with the NH of adjacent molecules, as shown in Figure 1. The hydrogen-bonding parameters (Table 1) are well within the normal ranges. It should be noted that the present structure is quite different from previously reported structures of both metal complexes and protonated salts of **1** since, in the former case, coordination of the nitrogen lone pairs to the metal dominates, whereas, in the latter, protonation removes any opportunity for hydrogen bonding.

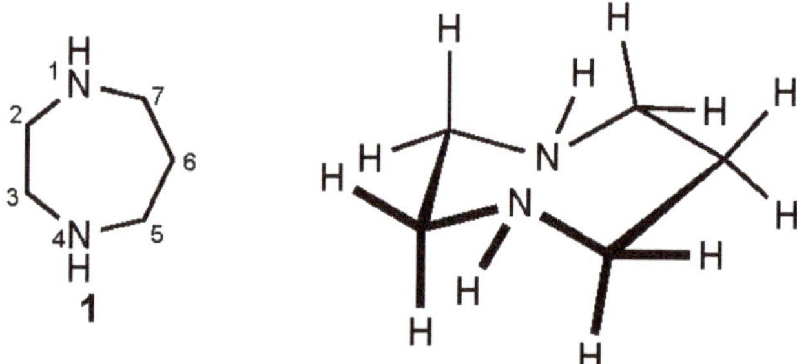

Figure 2. Structures of **1** showing IUPAC numbering and observed conformation.

Table 1. Hydrogen bonding parameters for **1** (Å, °).

D—H ... A	D—H	H ... A	D ... A	D—H ... A
N(1)–H(1) ... N(1B)	0.977(7)	2.21(3)	3.189(4)	175(4)

The symmetry operator to generate N(1B) is Y + 1, −X + 1, −Z + 1.

The ^{1}H and ^{13}C NMR spectra for **1** are documented in D$_2$O in a recent publication [4] along with HSQC and HMBC spectra, allowing complete and unambiguous assignment.

Our data in $CDCl_3$ were in good agreement with these values but, by recording the non-[1]H-decoupled ^{13}C spectrum, all three values of $^1J_{C-H}$ were readily determined as ^1J C(2)–H(2) 132.6 Hz, ^1J C(5)–H(5) 132.7 Hz, ^1J C(6)–H(6) 124.3 Hz, and, in addition, ^2J C(6)–H(5/7) could be measured as 2.6 Hz. The remaining $^2J_{C-H}$ values as well as longer-range couplings were not readily determined due to the complexity of the signals. The larger values of $^1J_{C-H}$ for CH_2N are in agreement with literature reports with values of 136.1 Hz for C-3 of a tetrahydroisoquinoline [34] and 132.0–139.0 Hz for C-4/5 of imidazolidines [35], while the value for C(6)–H(6) is in good agreement with 124.6 Hz for cyclohexane [36].

Finally, the ^{15}N NMR signal observed at −344.6 ppm with respect to $MeNO_2$ is in good agreement with those for similar cyclic secondary amines such as piperidine (−342.3) and pyrrolidine (−342.6) [37].

In summary, we obtained the X-ray crystal structure of homopiperazine for the first time and found the molecule to exist in a pseudo-chair conformation, with each NH acting both as a hydrogen-bonding donor and acceptor, leading to a complex network structure overall. The values of the three one-bond CH coupling constants and the ^{15}N chemical shift have been determined from its NMR spectra.

3. Experimental

The structure was determined on a Rigaku XtalLAB P200 diffractometer using graphite monochromated Mo Kα radiation $\lambda = 0.71075$ Å.

Crystal data for $C_5H_{12}N_2$, $M = 100.16$ g mol^{-1}, colourless prism, crystal dimensions $0.10 \times 0.10 \times 0.10$ mm^3, tetragonal, space group I–42d (No. 122), $a = b = 7.208(2)$, $c = 23.094(7)$ Å, $\alpha = \beta = \gamma = 90.00°$, $V = 1199.9(6)$ Å^3, $Z = 8$, $D_{calc} = 1.109$ g cm^{-3}, $T = 93$ K, $R1 = 0.0637$, $Rw2 = 0.1633$ for 536 reflections with $I > 2\sigma(I)$, and 37 variables, R_{int} 0.0374, goodness of fit on F^2 1.137. Data have been deposited at the Cambridge Crystallographic Data Centre as CCDC 2049324. The data can be obtained free of charge from the Cambridge Crystallographic Data Centre via http://www.ccdc.cam.ac.uk/getstructures (accessed on 30 March 2021). The structure was solved by direct methods and refined by full-matrix least-squares against F^2 (SHELXL Version 2018/3 [38]).

NMR spectra were recorded using a Bruker (Bruker, Billerica, MA, USA) AV instrument at 75.458 MHz (^{13}C) and a Bruker AVIII-HD instrument at 50.69 MHz (^{15}N) in $CDCl_3$ with chemical shifts given with respect to Me_4Si (^{13}C) and $MeNO_2$ (^{15}N) and coupling constants in Hz.

δ_C (non-decoupled) 33.3 (t of quintets, ^1J 124.3, ^2J 2.6, C-6), 47.8 (t of m, ^1J 132.7, C-5,7), 51.5 (t of m, ^1J 132.6, C-2,3); δ_N −344.6.

Supplementary Materials: The following are available online, Figure S1: Undecoupled ^{13}C-NMR spectrum of **1**; Figure S2: Undecoupled ^{13}C-NMR spectrum of **1**—expansion of C-6 signal; Figure S3: ^{15}N-NMR spectrum of **1**.

Author Contributions: A.M.Z.S. collected the X-ray data; D.K.S. obtained a preliminary solution of the structure, which was then optimised by A.M.Z.S.; R.A.A. designed the experiments, ran the NMR spectra, analysed the data, and wrote the paper. All authors have read and agreed to the published version of the manuscript.

Funding: This research received no external funding.

Data Availability Statement: The X-ray data are at CCDC as stated in the paper, and all NMR data are in the Supplementary Materials.

Conflicts of Interest: The authors declare no conflict of interest.

References

1. Bleier, L. Bildung secundärer Basen aus Aethylendiamin. *Ber. Dtsch. Chem. Ges.* **1899**, *32*, 1825–1830. [CrossRef]
2. Howard, C.C.; Marckwald, W. Ueber das Bistrimethylendiimin. *Ber. Dtsch. Chem. Ges.* **1899**, *32*, 2038–2042. [CrossRef]
3. Poppelsdorf, F.; Myerly, R.C. A novel synthesis of homopiperazine and its monomethyl derivatives. *J. Org. Chem.* **1961**, *26*, 131–134. [CrossRef]

4. Kim, Y.; Choi, J.; Nam, S.; Jeong, S.; Yoon, Y. NMR study of carbon dioxide absorption in aqueous potassium carbonate and homopiperazine blend. *Energy Fuels* **2012**, *26*, 1449–1458. [CrossRef]

5. Liang, X.; Chen, Y.; Wang, L.; Zhang, F.; Fan, Z.; Cao, T.; Cao, Y.; Zhu, H.; He, X.; Deng, B.; et al. Effect of carbon-skeleton isomerism on the dielectric properties and proton conduction of organic cocrystal compounds assembled from 1,2,4,5-benzenetetracarboxylic acid and piperazine derivatives. *New J. Chem.* **2019**, *43*, 11099–11112. [CrossRef]

6. Du, Y.; Li, H.; Wang, Z.; Zhang, M.; Liu, K.; Liu, Y.; Chen, R.; Wang, L. Supramolecular assemblies of bi-component molecular solids formed between homopiperazine and organic acids. *J. Mol. Struct.* **2019**, *1196*, 828–835. [CrossRef]

7. Guo, W.; Fu, X.; Chen, J. Supramolecular adducts of mesocyclic diamines with various carboxylic acids: Charge-assisted hydrogen-bonding in molecular recognition. *J. Saudi Chem. Soc.* **2020**, *24*, 885–895. [CrossRef]

8. Sun, Y.-L.; Shi, C.; Zhang, W. Distinct room-temperature dielectric transition in a perchlorate-based organic-inorganic hybrid perovskite. *Dalton Trans.* **2017**, *46*, 16774–16778. [CrossRef] [PubMed]

9. Shang, Y.; Huang, R.-K.; Chen, S.-L.; He, C.-T.; Yu, Z.-H.; Ye, Y.-M.; Zhang, W.-X.; Chen, X.-M. Metal-free molecular perovskite high-energetic materials. *Cryst. Growth Des.* **2020**, *20*, 1891–1897. [CrossRef]

10. Li, T.; Chen, C.; Guo, F.; Li, J.; Zeng, H.; Lin, Z. Extra-large-pore metal sulfate-oxalates with diamondoid and zeolitic frameworks. *Inorg. Chem. Commun.* **2018**, *93*, 33–36. [CrossRef]

11. Sahbani, T.; Sta, W.S.; Al-Deyab, S.S.; Rzaigui, M. Homopiperazine-1,4-diium bis[hexaaquacobalt(II)] trisulfate. *Acta Crystallogr. Sect. E* **2011**, *67*, m1079. [CrossRef]

12. Natarajan, S. Hydro/solvothermal synthesis and structures of new zinc phosphates of varying dimensionality. *Inorg. Chem.* **2002**, *41*, 5530–5537. [CrossRef]

13. Wilkinson, H.S.; Harrison, W.T.A. Homopiperazinium bis(dihydrogenarsenate). *Acta Crystallogr. Sect. E* **2006**, *62*, m1397–m1399. [CrossRef]

14. Hemissi, H.; Rzaigui, M.; Al-Deyab, S.S. Bis(homopiperazine-1,4-diium) cyclotetraphosphate-telluric acid (1/2). *Acta Crystallogr. Sect. E* **2010**, *66*, o2712. [CrossRef] [PubMed]

15. Ameur, I.; Abid, S.; Besbes-Hentati, S.; Al-Deyab, S.S.; Rzaigui, M. Structural, vibrational, thermal and electrochemical studies of a cyclohexaphosphate complex, $(C_5H_{14}N_2)_2Cd_2Cl_2P_6O_{18} \cdot 4H_2O$. *Phosphorus Sulfur Silicon Relat. Elem.* **2013**, *188*, 1703–1712. [CrossRef]

16. Dhieb, A.C.; Dridi, I.; Mathlouthi, M.; Rzaigui, M.; Smirani, W. Structural physico chemical studies and biological analyses of a cadmium cluster complex. *J. Cluster Sci.* **2018**, *29*, 1123–1131. [CrossRef]

17. Said, M.; Boughzala, H. Structural characterization and physicochemical features of new coordination polymer homopiperazine-1,4-diium tetrachlorocadmate(II). *J. Mol. Struct.* **2020**, *1220*, 128696. [CrossRef]

18. Mao, L.; Guo, P.; Kepenekian, M.; Hadar, I.; Katan, C.; Even, J.; Schaller, R.D.; Stoumpos, C.C.; Kanatzidis, M.G. Structural diversity in white-light-emitting hybrid lead bromide perovskites. *J. Am. Chem. Soc.* **2018**, *140*, 13078–13088. [CrossRef]

19. Shestimerova, T.A.; Mironov, A.V.; Bykov, M.A.; Grigorieva, A.V.; Wei, Z.; Dikarev, E.V.; Shevelkov, A.V. Assembling polyiodides and iodobismuthates using a template effect of a cyclic diammonium cation and formation of a low-gap hybrid iodobismuthate with high thermal stability. *Molecules* **2020**, *25*, 2765. [CrossRef]

20. Almond, P.M.; Deakin, L.; Mar, A.; Albrecht-Schmitt, T.E. Hydrothermal syntheses, structures, and magnetic properties of the U(IV) fluorides $(C_5H_{14}N_2)_2U_2F_{12} \cdot 5H_2O$ and $(NH_4)_7U_6F_{31}$. *J. Solid State Chem.* **2001**, *158*, 87–93. [CrossRef]

21. Doran, M.B.; Norquist, A.J.; O'Hare, D. Exploration of composition space in templated uranium sulfates. *Inorg. Chem.* **2003**, *42*, 6989–6995. [CrossRef] [PubMed]

22. Guo, Y.-M.; Du, M.; Bu, X.-H. Bis(1,4-diazacycloheptane-*N,N'*)nickel(II) perchlorate. *Acta Crystallogr. Sect. E* **2001**, *57*, m280–m282. [CrossRef]

23. Klai, K.; Kaabi, K.; Jelsch, C.; Wenger, E.; Lefebvre, F.; Ben Nasr, C. A Hirshfeld surface analysis, synthesis, structure and characterization of a new Ni(II) diamagnetic complex with the bidentate ligand homopiperazine. *J. Mol. Struct.* **2017**, *1148*, 412–420. [CrossRef]

24. Xing, Z.; Yin, H.-B. Crystal structure of (1,4-diazepane)$_4$Cu$^{II}_2$(μ-Cl)$_{10}$Cu$^I_6$, $C_{20}H_{48}Cl_{10}Cu_8N_8$. *Z. Kristallogr. NCS* **2019**, *243*, 391–392. [CrossRef]

25. Ling, E.C.H.; Allen, G.W.; Hambley, T.W. DNA binding of a platinum(II) complex designed to bind interstrand but not intrastrand. *J. Am. Chem. Soc.* **1994**, *116*, 2673–2674. [CrossRef]

26. Ali, M.S.; Powers, C.A.; Whitmire, K.H.; Guzman-Jimenez, I.; Khokhar, A.R. Synthesis, characterization, and representative crystal structure of lipophilic platinum(II) (homopiperazine)carboxylate complexes. *J. Coord. Chem.* **2001**, *52*, 273–287. [CrossRef]

27. Ali, M.S.; Mukhopadhyay, U.; Shirvani, S.M.; Thurston, J.; Whitmire, K.H.; Khokhar, A.R. Homopiperazine platinum(II) complexes containing substituted disulfide groups: Crystal structure of [Pt^{II}(homopiperazine)(diphenylsulfide)Cl]NO$_3$. *Polyhedron* **2002**, *21*, 125–131. [CrossRef]

28. Ali, M.S.; Longoria, E., Jr.; Ely, T.O.; Whitmire, K.H.; Khokhar, A.R. Homopiperazine Pt(II) adducts with DNA bases and nucleosides: Crystal structure of [Pt^{II}(homopiperazine)(9-ethylguanine)$_2$](NO$_3$)$_2$. *Polyhedron* **2006**, *25*, 2065–2071. [CrossRef]

29. Allen, F.H.; Kennard, O.; Watson, D.G.; Brammer, L.; Orpen, A.G.; Taylor, R. Tables of bond lengths determined by X-ray and neutron diffraction. Part 1. Bond lengths in organic compounds. *J. Chem. Soc. Perkin Trans. 2* **1987**, S1–S19. [CrossRef]

30. Fonari, M.S.; Antal, S.; Castañeda, R.; Ordonez, C.; Timofeeva, T.V. Crystalline products of CO$_2$ capture by piperazine aqueous solutions. *CrystEngComm* **2016**, *18*, 6282–6289. [CrossRef]

31. Sui, Y.; Hu, R.-H.; Luo, Z.-G.; Wen, J.-W.; Zhong, L.-J. Crystal structure and switchable dielectric properties of singly protonated homopiperazinium perchlorate. *Asian J. Chem.* **2015**, *27*, 2876–2878. [CrossRef]

32. Greiser, J.; Kühnel, C.; Görls, H.; Weigand, W.; Freesmeyer, M. *N*,1,4-Tri(4-alkoxy-2-hydroxybenzyl)-DAZA: Efficient one-pot synthesis and labelling with ^{68}Ga for PET liver imaging *in ovo*. *Dalton Trans.* **2018**, *47*, 9000–9007. [CrossRef]

33. Coyle, J.; Downard, A.J.; Nelson, J.; McKee, V.; Harding, C.J.; Herbst-Irmer, R. Electrochemistry of a labile average-valence dicopper system. *Dalton Trans.* **2004**, *15*, 2357–2363. [CrossRef]

34. Mohamed, Y.A.H.; Chang, C.-J.; McLaughlin, J.L. Cactus alkaloids. XXXIX. A glucotetrahydroisoquinoline from the Mexican cactus *Pterocereus gaumeri*. *J. Nat. Prod.* **1979**, *42*, 197–202. [CrossRef] [PubMed]

35. Albrand, J.P.; Cogne, A.; Gagnaire, D.; Robert, J.B. Analyse des spectres de RMN d'imidazolidines. Utilisation des satellites ^{13}C. *Tetrahedron* **1971**, *27*, 2453–2461. [CrossRef]

36. Chertkov, V.A.; Sergeyev, N.M. ^{13}C–^{1}H coupling constants in cyclohexane. *J. Am. Chem. Soc.* **1977**, *99*, 6750–6752. [CrossRef]

37. Dokalik, A.; Kalchhauser, H.; Mikenda, W.; Schweng, G. NMR spectra of nitrogen-containing compounds. Correlations between experimental and GIAO calculated data. *Magn. Reson. Chem.* **1999**, *37*, 895–902. [CrossRef]

38. Sheldrick, G.M. A short history of SHELXL. *Acta Crystallogr. Sect. A* **2008**, *64*, 112–122. [CrossRef]

MDPI
St. Alban-Anlage 66
4052 Basel
Switzerland
Tel. +41 61 683 77 34
Fax +41 61 302 89 18
www.mdpi.com

Molbank Editorial Office
E-mail: molbank@mdpi.com
www.mdpi.com/journal/molbank